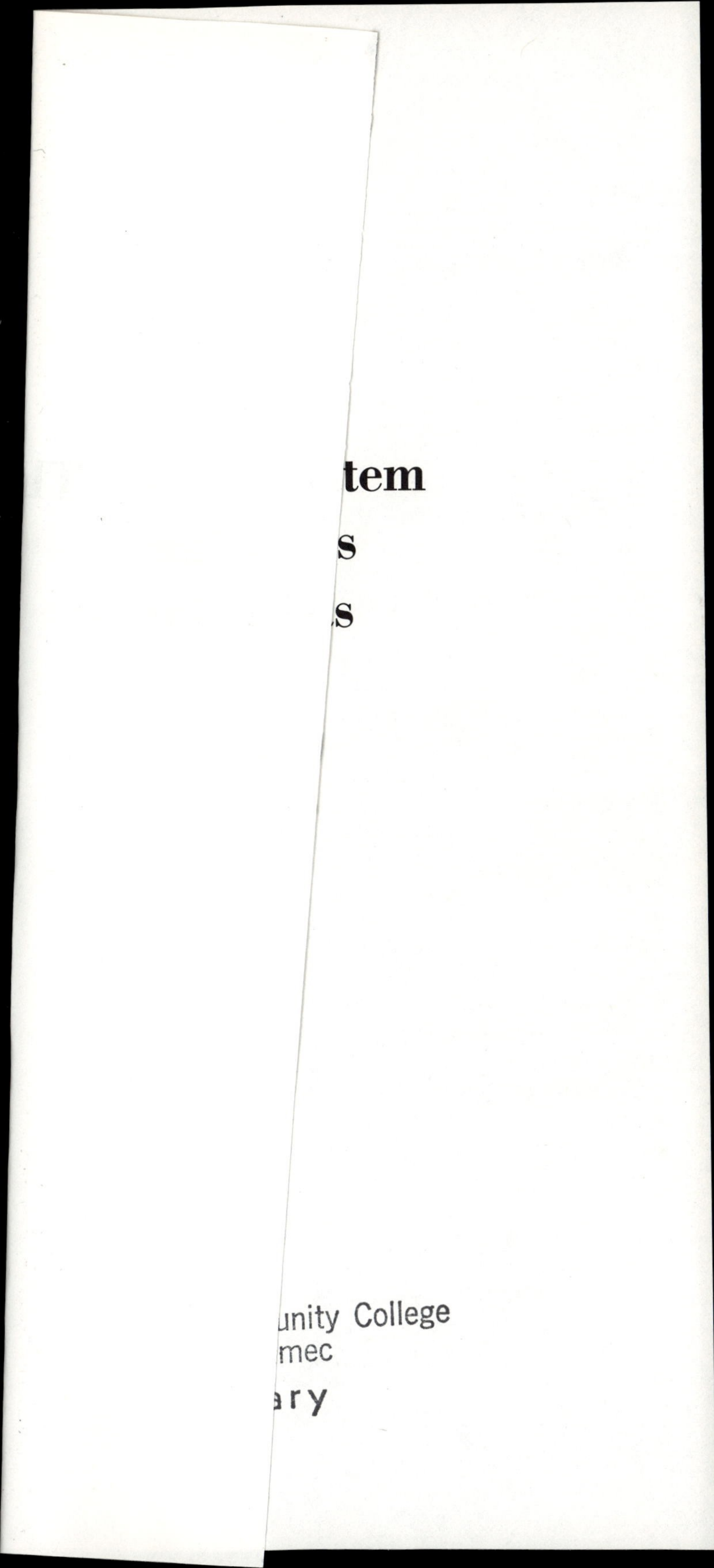

tem

s

s

The Writing System for Engineers and Scientists

Edmond H. Weiss

Prentice-Hall, Inc., Englewood Cliffs, New Jersey
07632

Library of Congress Cataloging in Publication Data

WEISS, EDMOND H.
The writing system for engineers and scientists.

Bibliography: p.
Includes index.
1. Technical writing. I. Title.
T11.W44 808.06662021 81-775
ISBN 0-13-971606-8 AACR2

Editorial/production supervision
and interior design by Anne Bridgman
Cover design by Dawn L. Stanley
Manufacturing buyer: Ed O'Dougherty

Printed in the United States of America

10 9 8 7 6 5 4 3 2 1

Prentice-Hall International, Inc., *London*
Prentice-Hall of Australia Pty. Limited, *Sydney*
Prentice-Hall of Canada, Ltd., *Toronto*
Prentice-Hall of India Private Limited, *New Delhi*
Prentice-Hall of Japan, Inc., *Tokyo*
Prentice-Hall of Southeast Asia Pte. Ltd., *Singapore*
Whitehall Books Limited, *Wellington, New Zealand*

For Joe Chapline

Contents

II

PLANNING, 23

III

DESIGN GUIDELINES, 71

Preface

Every sentence you write runs the risk of being read. So, every sentence you write has the power to help you or hurt you. If you write better, you will probably get more interesting assignments and make more money. But if you write badly most of the time, you will fail repeatedly to reach your professional and financial goals.

Good writing leads to success—especially in the scientific and technological industries. Of course, we all know a few exceptions to this claim, brilliant scientists or engineers who can barely write a paragraph. But these rare exceptions hardly affect the rule: To be a competent engineer or scientist you must write competently; to be an excellent engineer or scientist you must write *well*.

THE WRITING SYSTEM FOR ENGINEERS AND SCIENTISTS has three main goals:

> First, to convince you that writing is a *fundamental* part of your work as an engineer or scientist—not just an incidental part or occasional nuisance; that until you take your writing seriously and resolve to write better you will not be as effective in your work as you should be.

Second, to prove to you that the skills needed to write effective reports, proposals, or manuals are the *same* skills needed to solve engineering or scientific problems; that your training and experience make you well equipped to be a good writer.

Third, to teach you a practical, feasible procedure to follow when you write; a *system* that will, on the one hand, increase the chances that your writing will be effective and, on the other, decrease the chances that you will waste your time and energy when you write.

THE WRITING SYSTEM FOR ENGINEERS AND SCIENTISTS, then, is not like most other books about technical or business writing. Granted, it does cover some of the same territory as the other books: how to outline a report; how to untangle an awkward sentence; how to trim away excess words.

What makes THE WRITING SYSTEM . . . different from scores of other books, though, is its emphasis on *strategy.* In this system (for which the book is named), writing is treated as a purposeful activity, a means to some end, in which the readers will know something, believe something, or do something that the writer wants. The process of writing is always part of some larger process; the objective of your proposal or memo or report is always part of some larger objective.

From this perspective, THE WRITING SYSTEM . . . argues that everything you write must be planned as carefully as any engineering or scientific project you begin. Your writing must start with a specification or intended outcome, and then you must *design* a written product that will meet the specification or produce the outcome. Writing without such a well-considered plan is not only the greatest single waste of a writer's time but also the greatest single cause of weak, unclear, ineffective communication.

The test of good writing, then, is whether it meets the writer's objective—*your* objective—and whether it advances the larger process of which it is a part. The test of the proposal is the approval or sale. The test of the manual is its usefulness to the operator. The test of your letter of complaint is whether the problem is adjusted. The test of your clinical paper is whether it is judged publishable by competent referees. And the test of your final report is whether your recommendations are actually followed.

This strategic view of writing—as obvious and basic as it may seem to some readers—is by no means universally accepted or embraced by engineers and scientists. Most of the engineers and scientists I meet in my seminars reject this approach—at least at first. Instead, they labor under one of two misconceptions.

First, many talk about their writing as though it were a purposeless—even counterproductive—activity, an irritating interruption of their "real" work, a low-level and inconsequential task that ought to be relegated to some semi-skilled workers called "writers." As one fellow put it to me, "I just wish they would let us chemists do chemistry and give the writing to writers."

Still others, though they do not regard their writing as purposeless, mistakenly see it as having only one legitimate purpose: to inform. As they put it, engineers and scientists must write only about facts; their sole aim is to transport hard data from one place to another. Those other functions of communication —persuading, inspiring, influencing decisions, selling—are not, in their view, the proper concerns of engineers and scientists.

Both groups are wrong. Good writing is integral to the progress of engineering and science—not to mention that of engineers and scientists. Technical work just cannot be considered complete until it has been written up and the write-ups have been shown to competent reviewers. And scientific ideas or findings cannot be judged sound, correct, or useful until they have been communicated, intelligibly, to the right scientific or business or government audience.

Further, the notion that writing about science and technology deals exclusively with facts is arbitrary and naive. Problems are not facts; conclusions are not facts; recommendations are not facts. Even when they are based on hard fact, the problems, conclusions, and recommendations must be put into a convincing case or a moving presentation. To be effective in their jobs, engineers and scientists must do much more than inform; they must explain, prove, evaluate, justify, defend, attack, choose, advocate, and refute. Facts alone can do none of these. Clear, well-argued reports and proposals can.

Does this mean that THE WRITING SYSTEM . . ., because of its emphasis on strategy, does not address the more basic and familiar writing topics: grammar, sentence structure, diction, and all the other demons that haunt so many writers?

Quite the contrary. THE WRITING SYSTEM . . . treats all these issues of language and "style" in their proper place. For, if everything you write has a purpose, and if every purpose calls for some response or reaction from the readers, then an effective writer must never do anything that will render the written product inaccessible or incomprehensible to those readers. Improvements in style, therefore, are essential for making the preliminary version of the written product (the "first draft") good enough to send to the readers.

Style is a practical issue, rather than a frustrating quest to meet some ideal standards. Good style—to put it plainly—makes your writing easier to read and less boring. Every time you improve your style you increase the chances that you will have the effect you want to have on your readers. Simplifying a sentence may double the number of people who can read and understand that sentence; changing *utilize* to *use* or *prioritize* to *rank* can make a tedious passage interesting.

Good style comes only from good editing. Here, again, THE WRITING SYSTEM . . . takes a position different from that of most books about writing —which seem to claim that a careful writer can compose clear, clean, readable sentences in a first draft. Indeed, many of the engineers and scientists I meet seem to share this unrealistic notion and condemn themselves for needing to edit their work. They talk as though this were wasteful and inefficient!

This book presumes that *no one* can write a first draft that is good enough to be the last draft. No one: neither amateur nor professional. Moreover, no one should try, and certainly no one should be embarrassed by trying and failing.

First drafts are just different from final drafts. Composing a first draft, deciding what you want to say and struggling to cast it in sentences and paragraphs, differs from the process of preparing that rough draft for a reader. First drafts are *by nature* long-winded, clumsy, inelegant, unclear. Trying to correct a first draft while you are first composing it is unnatural and will not work. It will merely add time.

Time. The desire to spend less time on writing is, ironically, what gets most engineers and scientists into trouble. They write without a plan—to save time. They try to edit their first draft while they are still composing it—to save time. They hurry the proofreading and never double-check the corrections—to save time.

These apparent timesavers are really time traps. They end up costing countless extra hours and days, making the task of writing all the less pleasant and productive. They not only increase the time needed to write but also make the effort onerous, what the prisoner calls "bad time." Indeed, about half the people who take my writing seminars name *time* as their biggest writing problem. They want to contain the effort, cut the job down to size, free themselves to spend more hours on their "real" work.

Will the system in this book save you time? Yes! Most engineers and scientists waste time when they write, and the procedures in this system were devised to prevent all of the most common time-wasters. So, if you are among the majority who waste time, then this book will teach you to save it.

This book is the result of about six years' worth of seminars taken by thousands of engineers and scientists, managers, computer programmers, and businesspeople. Although I have given the "same" two-day seminar hundreds of times, no two of them have really been the same. Every time I conduct the seminar I learn something new—about the strategy of reports or the vagaries of punctuation or some other aspect of the endlessly fascinating problem of clear writing.

More important, though, at each seminar I have learned more about the work of engineers and scientists; I have gathered firsthand information about the problems and pressures of their jobs, the conditions under which they must actually write their reports and memos. Conditions never even anticipated by their teachers of "communication" or "technical writing." In some ways, I learned as much as I taught. Now, I am offering some of what I have learned to you.

E.H.W.

THE SYSTEM

1

Effective Writing: A Systems Approach

EVERYTHING WE WRITE IS A PRODUCT

I once asked a client how he could tell when he had written a good report. He answered, "When I never hear another word about it ever again!" His answer was typical . . . and sad.

For many engineers and scientists, writing is the most tedious and frustrating part of their work. Starting a report is agony; editing it is torture; and rewriting it—over and over again—is humiliation. For these writers, then, a good report is one that goes away and does not bother the author anymore.

Unfortunately, though, people who feel this way about writing will probably never be very good at it. And thinking this way will almost guarantee that each memo and report will need to be revised several times. As understandable as the attitude may be, it nevertheless aggravates the problem.

The far more useful attitude is to think of each piece of writing as a product —an engineered, manufactured product that is supposed to perform some function or achieve some purpose. If we think of our reports and memos this way, then the answer to the question changes dramatically. A good report is no longer one that goes away without a trace. Rather, a good report (or proposal or article or manual) is one that does *precisely what it was designed to do.* The writers who think this way, instead of hoping for silence at the other end of the communication channel, will eagerly seek out the responses and reactions of their readers. They will want to know if the product worked!

For a piece of writing to succeed, then, its author must have some clear purpose or function in mind. The piece must be intended to *do* something to a reader (or group of readers). A written product—just like an aircraft brake, or a computer program that generates financial reports, or a technique used to evaluate new employees, or any other engineered product—must meet specifications and performance standards. And it is just as silly to write a report without these specifications and standards in mind as it is to, say, improvise the construction of a bridge. (Well, not quite that silly! But you get the idea.)

Written products, therefore, can fail in all the ways that other products can fail. A badly made report might do nothing at all because it is so clumsy in style and forbidding in appearance. (There is no way of knowing for sure what proportion of the documents written in American industry and government is never really read by anyone—but we know it is high.)

Some written products can perform the wrong function or invite misuse. Instead of comforting the readers, they alarm them. Instead of instructing the readers, they confuse them. And, in some horrible cases, instead of winning the assent and approval of the readers, they bring scorn and derision to the writer.

An incidental but important effect of all poor writing is that it harms the reputation of the writer. Careless, awkward letters and reports create the impression that the writer is lazy, rude (carelessness in business dealings usually seems rude), or incompetent. Certain errors can even make the author seem ignorant and uneducated, despite his or her "advanced degree." (I, for one, find it difficult to trust the intelligence of anyone who says *irregardless.* And I'm nearly as uncomfortable with people who say *administrate* or who confuse *infer* with *imply.*)

In contrast, good writing makes the ideas of the author seem more lively and interesting. A researcher who is an excellent writer will be thought of as an excellent researcher; this researcher's ideas will be better received, and he or she will probably get more money to work with. Bad writing calls attention to the defects in the writer; good writing complements the quality of the writer's ideas.

A written product, then, can perform or fail. When it performs, or succeeds, it must do the job or carry out the function meant for it. Or it can fail by doing nothing, or by doing the wrong thing, or by bringing scorn and criticism to the writer.

PREVENTING FAILURES

Almost every adult—certainly every adult in a professional job—can write a bit. When bosses or consultants tell engineers or scientists that they "can't write," they are obviously wrong.

The problem is that we all make mistakes when we write. All of us, amateur and professional alike. To be a better writer, we do not need to learn how to write but, rather, how to eliminate or control the errors and flaws that come up at each stage of the process.

So, logically, an introduction to the Writing System must dwell on the negative: mistakes, errors, false assumptions, pitfalls, failures.

Writing fails, usually, because writers fail. The burden of success is on the author, not the reader. The fact that the director of marketing tossed your report aside because he could not find what he was looking for is *your* fault. The fact that technicians misused a tool because your manual was too hard for them to read is *your* fault. The fact that your proposal came in second in a fair competitive bidding, even though you were the best offeror, is *your* fault.

And even if what I am saying is false, even if the best writer in the world could not have prevented these failures, even if the readers in question are certified incompetents, even then . . . the burden is still on the writer. Alas, there is no other place to put it, and there is no other attitude that helps a writer to succeed.

Of course, sometimes there are extenuating circumstances for the failures of your reports and documents. Of course, some purposes are virtually impossible to achieve; some readers are too stupid to reach; some truths are too painful to accept.

Most failures, though, are the direct responsibility of the author: especially those failures that force the author to revise a report a dozen times.

Engineers and scientists, when they fail as authors, usually fail for one or more of these three reasons:

- **They do not plan** — believing that writing, unlike other technical processes, can be improvised; or
- **They do not read and edit carefully** — believing that what they have written is clear when, too often, it is awkward and unreadable . . . clear only to the author; or
- **They waste time** — trying too long to do the impossible while neglecting practical and important details.

Writers fail, therefore, not because they cannot write but because they neglect to do what needs to be done. Either because they do not know any better, or because they are searching for shortcuts, or because they believe there is not enough time in their day to do it the proper way, they neglect the fundamental requirements for effective writing. All of us, including professional writers, must plan before we write and must edit our first draft. And, even if skipping these tasks will occasionally appear to save time, it will actually not only degrade the quality of what we write but also waste more of our time than it saves!

Failures of Design: Strategic Errors

Every piece we write needs a plan or a design. This design includes an outline, but the outline is only part of the design. (In fact, it is more useful to think that the outline follows from the design, rather than being a part of it.)

The heart of the design is a simple, direct statement of the *subject* of the piece, the intended *audience* or (audiences), and a specific, precisely worded, *function* or *purpose* to be accomplished with that audience. Putting it bluntly, until authors have a definite idea of their subject, their audience, and especially their purpose, they are foolhardy even to write an outline, let alone a first draft.

For most engineers and scientists, defining the subject of a message is easy. They are, after all, subject-oriented people, experts in the subject they are writing about. But analyzing and defining the audience, and especially the purpose, are difficult for most technical specialists.

The results of failing to plan and design are *strategic errors,* errors that undermine the entire look and "feel" of the report or document, errors that cause your report to misfire or even blow up in your own hands.

As an example, consider this opening paragraph from a computer service company's *Customer Newsletter:*

Dear Omega Customer:

(1) For several months, Omega personnel have been devoting
(2) considerable effort toward the implementation of the IBM
(3) MVS (Multiple Virtual Storage) Operating System (OS/VS2
(4) Release 3.7) and JES2 (Job Entry Sybsystem). MVS and JES2
(5) are functional replacements for Omega's present SVS (Single
(6) Virtual Storage) Operating System and the HASP spooling
(7) system. Because MVS and JES2 are largely upward compatible
(8) with the present Omega operating softward, the transition is
(9) intended to be operationally transparent to Omega customers.

The obvious flaw in this passage is, of course, the obscure vocabulary, the use of terminology familiar to people in the computer industry but unfamiliar to most of the customers of the company. The first sentence, lines 1–4, even if it were understandable, fails as an opening; it says nothing that explains to the reader what the effect of the change (if any) will be; it congratulates the staff of the company (without describing any real attainments) and permits the reader to ask, So what?

The third sentence, on line 7, seems to promise that the new system will be "upward compatible" with the old (that is, compatible), and then it takes away the promise with the qualifier "largely." Not only is the expression "largely compatible" semantically absurd (like *almost pregnant*), it is a disastrous way to give confidence to the reader, whose interest suddenly shifts, with alarm, to the ways in which the new system is *not* compatible. (Incidentally, we are never told those ways.)

The final clause, lines 8 and 9, seems to promise that the customer will be unaffected by the change ("transparent" is computer talk for "invisible"), but a closer reading of the dense, pseudotechnical language in the last clause shows that there is no promise at all, merely an "intention." (And the use of the passive construction in the last clause leaves us uncertain about who is intending what. "The transition is intended" by whom?)

The passage is, by any criterion, a dismal failure. But not because it would displease an English teacher or a technical editor. It is a failure because it will not perform the function it was meant to perform: to alert the customers of the company to a change of service that will give them certain specific advantages at the cost of certain short-term inconveniences.

The best thing that can be said of this excerpt is that it demonstrates so beautifully the dangers of writing without a thoughtful design.

Failures of Editing: Tactical Errors

When most people think about writing faults, they are less likely to think of the strategic errors mentioned earlier than of the dozens of errors of usage, grammar, and style that fill most of the books on business and technical writing.

Everyone's writing needs editing for these *tactical errors,* mistakes in the selection of words and the construction of sentences. No one (yes, no one) can write a first draft good enough to be the last draft, and no one should even try.

Many authors refer to these tactical errors under the somewhat disparaging heading of "grammar" or "style" or "mechanics." Any of these names is acceptable as long as you realize that to be a good writer you must be skillful in these "mechanics." The practice in some schools of giving English themes two grades —one for substance and one for mechanics—probably does more harm than good. It is useless to try to separate the evaluation of an idea from the evaluation of the language in which it is expressed. Certainly an engineer could be brilliant at engineering and horrible at the tactics of writing—but it is nearly impossible for us, the engineer's readers, to know when this is the case.

You may refer to these tactical items as "mechanics," therefore, only in the same sense that you refer to "celestial mechanics." The mechanics of language are the laws of thought and communication.

And, remember that these tactical errors are the result of poor editing, not of poor writing. The point to be stressed is that all writers, including the very best, make scores of mistakes in their first drafts. But the successful writers catch and correct them before the final version.

No one can write a first draft free of stylistic and mechanical errors. No one should even try. But everyone is obliged to seek out these errors, to screen and filter them out, before the message reaches the intended audience.

Failures of Management: Excessive Time and Trouble

I have never met anyone who wanted to spend more time writing reports and proposals. It is slow, exhausting work, even for people who are good at it. (Most professional writers agree with Dorothy Parker that they hate writing and love having written.)

I also know that most of the writing time wasted by engineers and scientists is spent on their own incorrect writing habits. Specifically, writers waste time in the following six ways:

1. *They write things that do not need to be written at all.* Instead of handling a problem with a phone call or conversation, they write a letter. Instead of waiting long enough to get the facts about a case, they send a premature memo. They never stop for a moment to ask if this written message is necessary or desirable.
2. *They write without a plan.* Writing without a design guarantees a slow start, a weak opening, and at least one major revision—written after they finally decide what they are trying to accomplish.
3. *They neglect to get approval before the draft.* Because most documents written in science and industry need to be approved by the author's superior, it is wasteful to write an entire draft without having first secured approval of the plan and outline. The productive, efficient place to argue about the scope and contents of a written product is in the outlining stage, when major revisions can be made in a few seconds instead of a few days.
4. *They try to write a first draft good enough to be the last draft.* It takes three to four times longer to write a first draft that is meant to be the last draft than to write a first draft that is meant to be a first draft. And since no one's first draft is good enough to be sent or published anyway, the time is wasted. It is far more efficient to write a first draft without any regard for style, and then to edit and polish it later, than to edit it as it is being written.
5. *They try to edit all alone.* Most of the improvements to be made in a first draft should be made by the author. Ultimately, though, every writer who wants to be clear will also want to ask another reader (or professional editor) for help. Put simply, another reader can detect problems in a minute that the author could not detect in hours. Writers are too close to their own writing to know when it is unclear. (They, after all, understand it perfectly.) And there are some necessary improvements that they will *never* see without help.
6. *They submit illegible manuscripts.* The more problems that writers create for typists and artists, the more of their *own* time they waste. The more difficult it is to produce their own reports and proposals, the more of their time (not just secretarial time) these manuscripts will take. And, finally, the sloppier the manuscript, the greater the chance that there will be embarrassing or even devastating mistakes in the final version.

These six practices, then, can double or triple the time it takes to produce a written product without improving the effectiveness of that product at all.

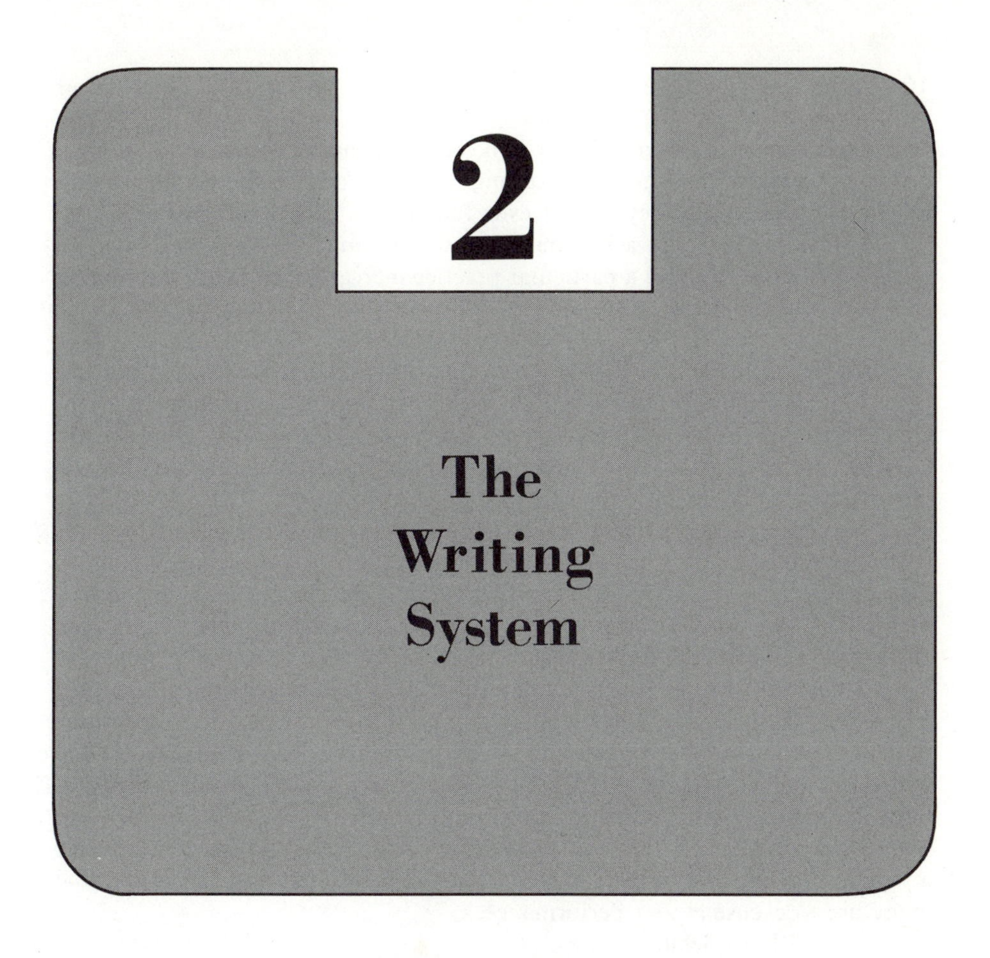

2 The Writing System

THE SOLUTION

The only way I know for anyone to become a better writer, that is, a writer who can find a way through this tangle of problems and defects, is to use a *systems approach.*

I am using the expression *systems approach* in its strict meaning. According to Russell Ackoff:

> In [the systems] approach a problem is not solved by taking it apart but by viewing it as a part of a larger problem. (*Redesigning the Future* [New York: Wiley-Interscience, 1974], p. 14.)

Thus, the key to becoming a better writer is to think of writing as part of a larger system, to view better writing as a means to some "higher end."

Depending on the circumstances, the function of your writing is either to help you achieve your own professional objectives or to help your organization achieve its organizational objectives. (Usually those ends are the same; in interesting cases they conflict and must be resolved.)

So, chemists must write better to be better chemists. Managers must write better to be better managers. Consulting engineers must write better because, unless they write well, they cannot even function as consulting engineers!

Every question you ask about your writing must be answered in terms of this "system end." Will a particular practice increase or decrease the chances that you will achieve your purpose, realize your professional or organizational aims?

Thus, writing in the passive voice (instead of the active) is bad not just because some English professor says so. Writing in the passive voice increases the chances that you will be misunderstood! Working from an outline is good not just because editors think so. A written outline increases the chances that you will meet your deadlines!

The systems approach, then, consists in reminding yourself periodically that everything you do when you are writing is related to some goal. When you have choices among several ways to write, you rate the choices according to how helpful they are to your purposes. When two ways seem equally helpful, you choose the easier or the less expensive (that is, less time-consuming) way.

To help writers, and to ensure that they take a systems approach to writing, I have devised a Writing System. As in any system, all the parts of the Writing System are related to each other and affect each other. There is a series of steps with decisions at each step, so that, at the end, writers can feel confident that they have written as "cost-effective" a message as possible.

Before you can use the Writing System, though, you must start to think like a writer. You must put aside some comfortable myths and misconceptions that many use to excuse a poor performance.

Ready? Let's debunk.

CORRECTING SOME MISCONCEPTIONS

To be a better writer, you must free yourself from several misconceptions about written products and the writing process. These self-defeating attitudes are, as often as not, an excuse for not doing better (really, not working harder) at the task.

First, contrary to what you may believe, *the difficulty in reading technical materials is due more to the weaknesses in the writing than to the difficulty or complexity of the technical content.* That you are writing about a difficult subject, even an esoteric or exotic subject, is no excuse for dense, difficult, unreadable text. Effective writers can explain almost anything to almost anyone—and the more difficult the subject, the harder they work at the writing. In contrast, many engineers and scientists, accustomed to writing only for readers with backgrounds like their own, will not bother to adapt their materials to their audience, assuming —incorrectly—that persons with the right training will be able to understand.

When you write badly, however, no one can understand you, no matter how expert. True, some people may be able to *guess* what you mean, but that is hardly the same as understanding you. Consider, for example, the following passage from a report about planning for the transportation of nuclear wastes:

> In *determining* the material flows, the nuclear industry was *characterized* by individual reactor projects located at specific sites. Site selection allows the direct interface of the *material flow projections* with the *transportation system models.* (Emphasis added.)

When I show this passage to engineers and scientists, they are usually baffled by it. If I ask them to edit or improve the passage, they plead ignorance: They do not know enough about the nuclear industry, they say, to interpret the passage.

Nonsense. The obscurity of this passage stems from its clumsy, dense writing, rather than from the complexity of its content. You do not have to know anything about nuclear technology to know that the words *determining* and *characterized* are misused in the passage. And you do not have to be a transportation specialist to know that such expressions as *material flow projections* and *transportation system models* are phrases that contain no grammar and are clear only to those people who already knew what the writer was thinking, even *before* they read it. To prove the point, consider the revised version:

> To better predict the flow of materials, we have depicted the nuclear industry as a set of individual reactor projects located at specific sites. Using specific sites, rather than regions, we can tie our projections of the flow of materials to the projections of change in the transportation system.

Notice that this revised version is not a "nontechnical" interpretation of the first. Rather, it is a clear, technical passage, suitable for its intended readers, no more difficult to understand than it needs to be!

A second misconception is that refinements of language and style do not count in business and technical communication. *Style always counts, because style demonstrates your competence and credibility.*

Writing that lacks style—writing that not only is clumsy and harder to read than it needs to be but also contains more errors than it should—irritates the reader. If your readers know something about writing, they will conclude that you are uneducated or incompetent.

Even if your readers know little about writing, however, they will still know that something is wrong. They will be distracted or confused, or they will have to spend longer with your memo or report than they want to. If your later writing does not improve, your readers will gradually consider you not only a poor writer but also an incompetent engineer or a confused scientist or an evasive contractor. Eventually your clumsy writing will be taken as evidence of clumsy thinking.

Of course, style counts more with people who do *not* know you than with people who do. Your colleagues and longtime associates will be more forgiving and will be less likely to judge you harshly because of an obscure paragraph or two.

Even within your inner circle, however, style counts for more than you realize. Because so much technical work culminates in a written product—a final report, a proposal, a manual, or a paper—your colleagues will come to resent the fact that *your* portions of the writing need rewriting more often than theirs. And if *they* are the ones who have to do the revising, their patience and tolerance will not last forever. A supervisor who must spend most of his or her time revising *your* writing is sure to remember that fact at personnel review time.

A third misconception is that good writing is a mysterious talent that cannot be taught or learned. *Most of what hurts our writing can be analyzed and corrected in a few hours.* Being a competent writer consists mainly in avoiding certain common, easily identified mistakes.

Interestingly, many engineers and scientists would like to believe that good writing will forever elude them, that it is a gift they lack and can never acquire. Often, a new student in one of my seminars will come up to me and say, "I are an engineer." Some of these people have been hiding behind that lame joke for twenty or thirty years.

Listen: Good writing is a technique. Competent writing can be learned by anyone with a basic grasp of the language—which is to say anyone who is skilled enough to earn a degree from an English-speaking college or university. (People who speak English as their second language usually—but not always—have to struggle with the mechanics of the language before they can become effective writers. Many *do* succeed.)

The simple truth is that in only a few hours you could learn to recognize the fifteen or twenty most common writing errors; in a few more you could learn to stop committing those errors. After just a few days, then, you could be among the upper third of writers—just that easily.

In two weeks you could learn the hundred or so most common tactical errors and how to avoid them; after that you would be in the upper tenth of writers!

Becoming a competent writer consists mainly in learning to avoid errors: errors of planning, errors of language and style, errors of process. Although overcoming these errors will not turn you into a brilliant or marvelous writer, it will, nevertheless, turn you into a competent engineer or scientist who can use the language to achieve his or her personal and professional objectives.

THE WRITING SYSTEM: AN OVERVIEW

The Writing System is a chain of consecutive procedures and tasks, interrupted frequently by small decisions and reviews. As in the simplest of computer programs, the process keeps moving forward as long as the decisions and approvals are favorable; if they are unfavorable, the process "loops" back to the previous step, which is repeated until the decision or approval allows the process to resume.

There are three main phases or components in the Writing System (as shown in the summary in Figure 2-1):

- **The planning stage,** in which the writers decide what (if anything) needs to be written and prepare a plan and outline
- **Attacking the draft,** in which the writers zealously generate a version that "covers" all the points in the plan and outline
- **Editing and revising,** in which the first version (or draft) is revised, refined, and eventually produced in final form

As you will see shortly, most of what is to be learned about better writing applies either to the planning component or to the editing component. Almost everything that writers do wrong is an error either of planning or of editing.

Yet, ironically, most of my clients and students think their problem is with the first draft; they have trouble getting it started and organized (poor planning), or they have trouble making it clear and readable (poor editing).

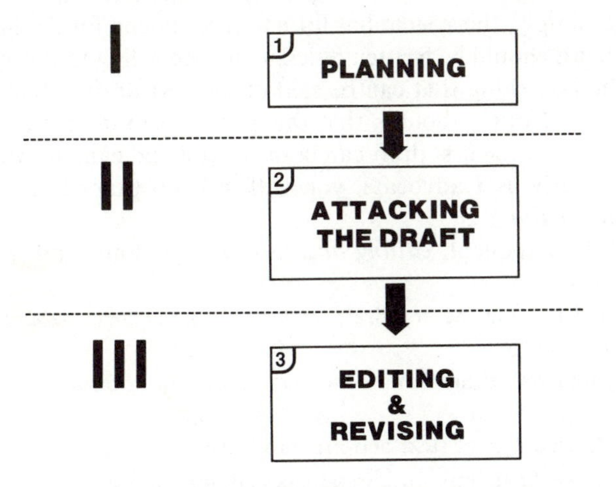

FIGURE 2-1 Overview of The Writing System

Put simply, you probably cannot write a coherent first draft without a plan, and you certainly (as you have now heard repeatedly) cannot write a first draft that is readable enough to be the last draft.

This system, even in the undetailed form you see here, should be an aid to every writer. If I can convince you that everything you write needs a *written plan* (with the possible exception of the one-sentence memo of transmittal) and that everything you write must be reread and revised before you send it out . . . if I can convince you of just that, even without telling you how to plan or edit, I will have improved your writing.

Obviously, though, the rest of this book is intended to describe what you should do in each component of the system, to give you as much sound advice as I can without actually standing over your shoulder and helping you to write.

Figure 2-2 shows the system at one lower level of detail. The first component, *planning,* actually has three main parts:

- A preliminary assessment of whether you really need to write at all (and I hope that occasionally you will decide *not* to)
- A formal planning activity that culminates in a written plan and outline
- A formal approval by the person who will have to accept or sign off the finished product—often the writer's supervisor or boss. Notice that this part of the planning process can "loop" until the supervisor approves the plan. (If no supervisor is involved, the writer personally approves the plan.)

The second component, *attacking the draft,* includes almost no additional detail because, simply, the system has little to recommend for the first draft other than noting that it should be written quickly and according to the approved plan. As I see it, the best thing that can be said of the first draft is that it is finished, ready for editing. Notice, though, that this phase also includes an "interval" in which the writer and the first draft can both cool off and calm down. (If you have written it as swiftly as I advocate, you will feel too excited and tired to edit. Editing requires *calm.*)

The third component, *editing and revising,* has four main parts:

- Editing and revising the first draft, making all the changes needed to enhance clarity and quality
- Securing approval, either from the writer's supervisor or from the writer personally
- Having the manuscript (text and art) produced
- Reviewing the final copy for correctness and appearance

The editing and revising component of the system is the most flexible; you may choose to do as little or as much editing and revising as you judge appropriate. (There is less free choice in deciding how long you will spend on the planning

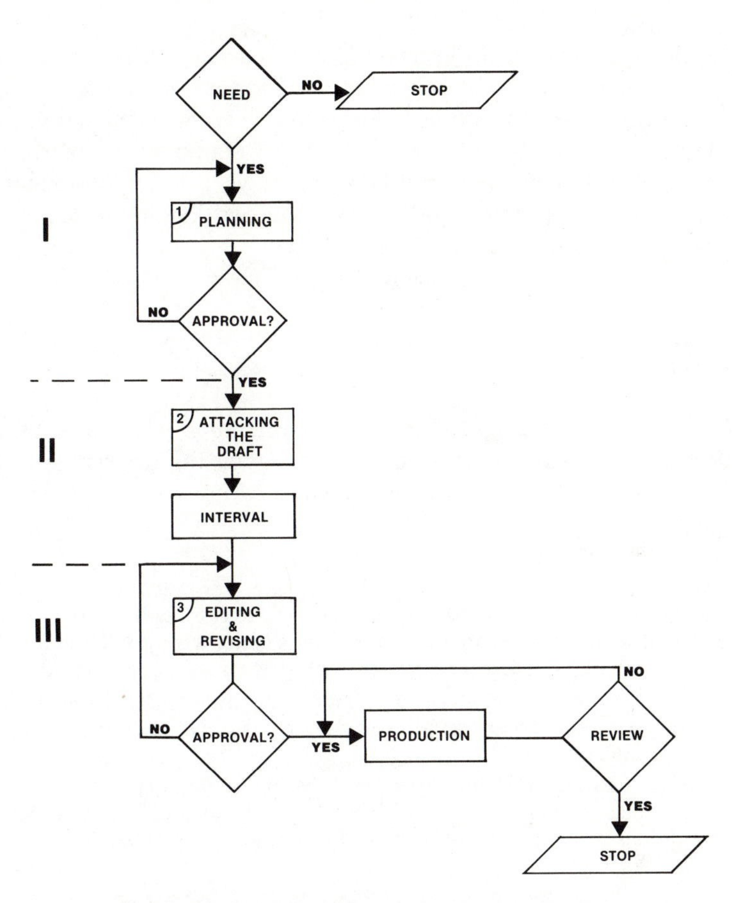

FIGURE 2-2 The Writing System: Level II

and first-draft stages; usually you will choose to spend as little time as you feel you need to get the job done.)

To complicate your choice, you should be aware that there is a direct relationship between the time you spend on the third component and the quality of the finished product. Of course, the relationship is not linear; five hours of editing will not give you five times as much quality as one hour. Still, you can keep editing and revising for quite a while before you reach the point of diminishing returns. You must decide how high to set the standards and then budget your editing time accordingly.

(Most books about writing are really books about editing. Every author who has enjoined you to write in shorter sentences and use more "active" verbs has really been telling you *how to improve your first draft.* Even many of the remarks I have made so far—complaining about *irregardless,* for example—are addressed to your editing, not to your first drafts.)

The reason that many writers are confused and overwhelmed by the job of writing is that they try to do everything at once rather than in a deliberate sequence. Consequently, I want you to understand that there is a time to worry about style and grammar and a time to forget it! If you know which time is which, your writing is more likely to be stylistically and grammatically correct.

The Writing System, then, can be regarded as a device that helps you to organize and apply what you already know about writing and to use everything you will be taught in the future.

In the sections that follow, each of the components is presented in somewhat greater detail.

The Planning Component

Good planning leads to the *right* written message, a report or letter that suits your own ends and is well adapted to your reader. As you can see in Figure 2-3, the process begins with a decision (whether or not to write) and ends with a formal, written plan:

1.1 **Assess the Need to Write.** Examine the impulses pressuring you to write. Is the need real? Will the impulse pass? Can you achieve your objective without writing? What are the consequences of not writing? (This task, therefore, eliminates the first time-waster: writing what need not be written at all.)

1.2 **Define the Subject and Occasion.** Name the project. Decide what you are writing about, and what occasioned the writing. Give the project a file location or address.

1.3 **Analyze the Function and Purpose.** Decide approximately what you are trying to accomplish. Do you wish to inform? Persuade? Move? Under what conditions will you be satisfied that your message is a success? What are your "specs"?

1.4 **Analyze Your Audience.** What individuals or groups are you trying to reach? Visualize your reader(s). Who are your primary and secondary audiences? Do the interests of your audiences conflict? What characteristics or preferences of your readers must you respect or negotiate?

1.5 **Define the Precise Communication Objective.** State exactly what your reader should know, believe, or do as a result of receiving your message. (Steps 1.2–1.5 eliminate the second time-waster: writing without a clear plan.)

1.6 **Define the Strategic Problems and Solutions.** Review your objective and your audience. Are there any barriers or problems that will make it hard to achieve that objective with that audience? If so, what communication techniques should you use to avert the obstacles?

1.7 **Select the Form and Scale.** Choose a format from those available, or devise a new one. Estimate the scope and length; distribute portions. Decide how "defensible" the document must be, that is, how resistant it must be to attack and refutation.

1.8 **Organize, Assign "Raw" Material.** Review the "raw" materials that you intend to "cover" or discuss in the piece you are about to write. If there are too many, organize them into clusters. Assign each cluster to a place in your message.

1.9 **Decide if this is a One-Person Writing Job.**

1.10 **Prepare a Simple Plan and Outline.** If this is a one-person job, write a simple one- or two-page plan and outline, including a brief production schedule.

1.11 **Approve the Plan.** Show the one-person plan to the supervisor (if appropriate). Revise until it is acceptable.

1.12 **Prepare a Detailed Modular Plan and Outline.** If several people are needed to prepare the message, prepare a detailed outline or "storyboard" plan. Make the plan "modular"; prepare a detailed production schedule and alert the production people.

1.13 **Approve the Plan.** Present the storyboard for review and approval; rework and revise it until acceptable. (Steps 1.11 and 1.13 eliminate the third main time-waster: writing without an approved plan.)

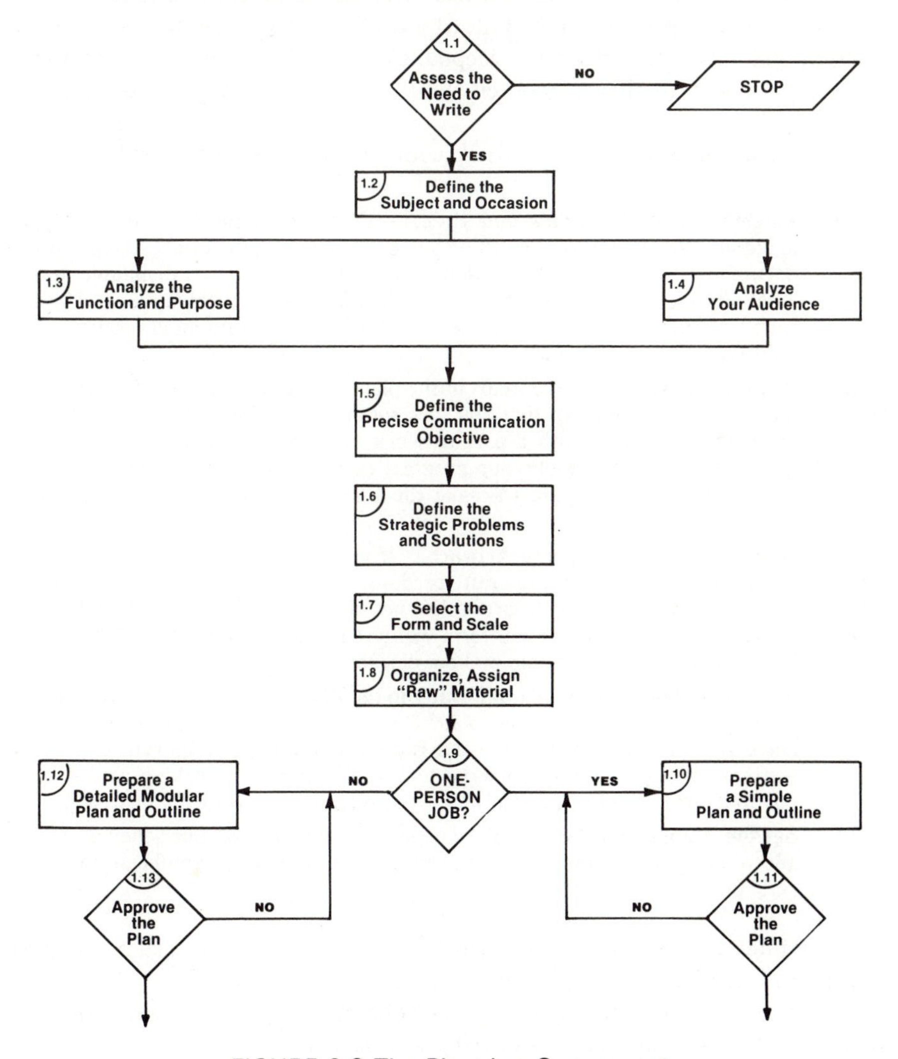

FIGURE 2-3 The Planning Component

So, at this stage in the process you have prepared either a short, one-person plan and outline (acceptable to you and your supervisor) or a detailed "storyboard" outline and detailed production schedule. In the first case, you are ready to write your first draft; in the second, you are ready—along with the other writers—to write your *parts* of the first draft.

Attacking the Draft

Given an approved plan and outline, you are set to write a hurried first draft.

To repeat, the Writing System has little advice on how to do that draft. Remember that you should not write the draft until you know your purpose precisely and have an approved or acceptable outline. Also remember that no one can write a first draft that is good enough to be the last draft—so do not even try.

There are five steps in the *attack* component (see Figure 2-4):

2.1 **Assemble Your Materials and Prepare Your Environment.** Put simply, make sure that you have everything you need (supplies, materials, equipment, documents, and so forth) and that you have a reasonably inhabitable and quiet place to work. Eliminate as many distractions and interruptions as you can (a big problem for many writers) and get yourself into an energetic and industrious frame of mind.

2.2 **Write Draft 1.** Cover the items in the plan and outline. If possible, write in the correct order (except that you save summaries for last). If you get stuck on a part, find some other part to work on. Write as much as you can; do not be neurotic about your progress or schedule. (This eliminates the fourth time-waster: trying to make the first draft good enough to be the last draft.)

2.3 **Do the Interim Typing and Artwork.** If your first-draft text is unreadable (or dictated), get it typed at draft speed—so that you can make revisions on a legible, typed text. If your artwork is not clear enough to proceed, and you are reasonably sure its present form will be suitable in the final version, have the artist or graphics people make a cleaner rendering.

2.4 **Review the Manuscript for Conformity to the Plan.** Have you covered every topic, theme, and issue in your plan? Have you followed your own strategy? Have you allocated the text in ways proportionate to the importance of the sections? If not, continue to write first-draft materials, working as fast as possible. Once the draft conforms to the plan, move on to the next step.

2.5 **Set the Manuscript Aside.** Take as long an interval as your schedule can afford and put the draft aside. Shift to a lower gear before you begin to edit and revise.

Take this interval seriously. Authors who try to revise during the heat of composition usually just have to revise again at a later time.

If your schedule will only allow a short break, try going for a walk or a run to ease your tension and relax your editorial nerves.

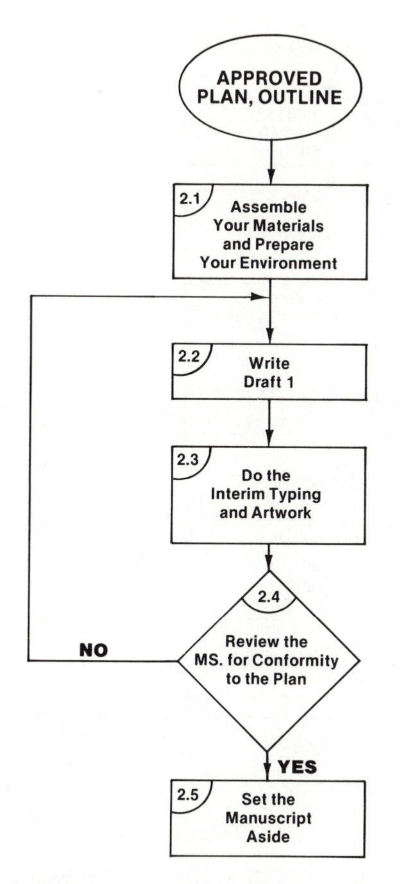

FIGURE 2-4 Attacking The Draft

Editing and Revising

As I said before, this third component, *editing and revising* (Figure 2-5), can take as long as you decide to spend on it. Usually, the time you have saved by hurrying through the first draft (rather than trying to make it the last draft) will be more than the time you need to edit and revise.

(If you are one of those writers who only writes first drafts and never revises, you will have to spend more time on your writing than before. The time will be well spent, though. Your unrevised first drafts are probably horrible.)

3.1 **Edit for Conformity to Strategy.** Now that you have calmed down, repeat the review you did in Step 2.4. Check to make sure that the overall shape and format of the document turned out the way you intended. (If, in the course of writing, you deliberately changed your original plan, then your first draft should correspond to your revised plan.)

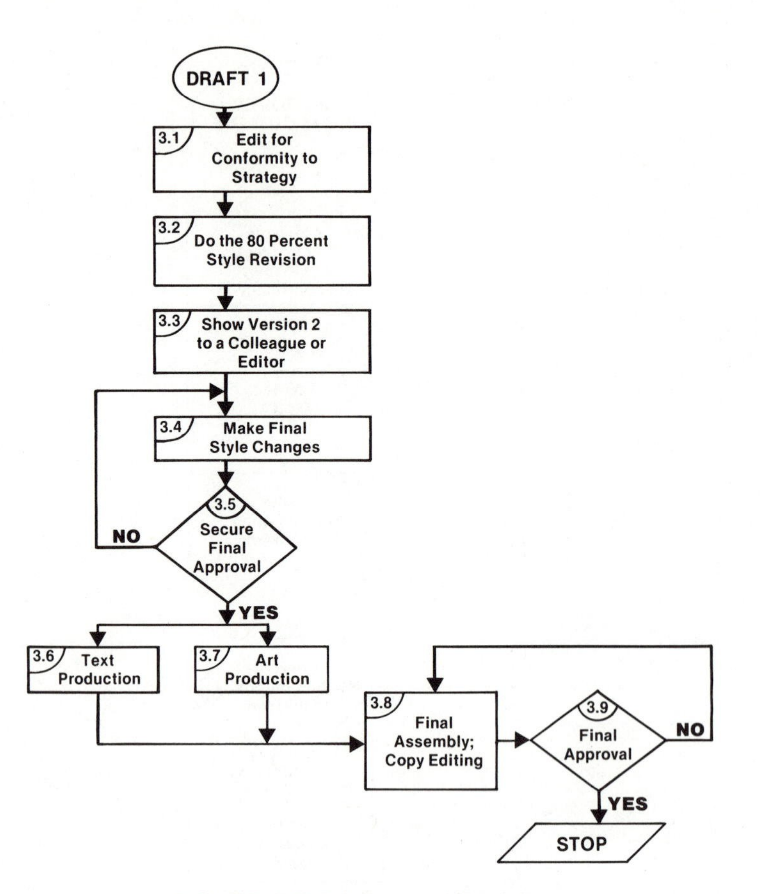

FIGURE 2-5 Editing and Revising

Add, cut, or rearrange sections until the draft has all the parts you intended to write, in the order you intended to present them.

3.2 **Do the 80 Percent Style Revision.** Make 80 percent of the improvements the first draft needs. Revise every word, phrase, or sentence that is in obvious need of improvement. Add some aids for the reader to make the draft more accessible. Make every editorial improvement that comes easily.

3.3 **Show Version 2 to a Colleague or Editor.** When there are no more obvious improvements to be made, show your revised draft to a colleague or editor (not your boss). Take advice on ways to make it clearer and more accessible. (Showing your 80 percent version to another person eliminates the fifth time-waster: trying to edit and write all alone.)

3.4 **Make Final Style Changes.** Follow the advice of your reader/editor, or, in some cases, allow that person to make the final improvements.

3.5 **Secure Final Approval.** Get your supervisor to approve your text and art; revise further until approval is secured.

3.6 **Text Production.** Send your text to typing, being sure that it is legible to the typists. (Sending an illegible manuscript to the typist is the sixth main time-waster.)

3.7 **Art Production.** Send your figures, exhibits, and tables to the appropriate people; communicate your desires clearly; explain to the artists where they may use artistic latitude.

3.8 **Final Assembly; Copy Editing.** Bring the parts together; ensure that the document is complete and in sequence. Study the document for mechanical errors. Appraise and improve the overall appearance. Supervise any necessary corrections. Correct all mechanical errors. Prepare the document for reproduction, if appropriate.

Notice that the system begins with large questions of function and purpose and ends with tiny questions of mechanics and typing. The process is anticlimactic, getting less and less interesting after the beginning. Even so, the small, fastidious details that are the heart of the last few steps can be essential to the overall impact of the message.

Certainly, no amount of good style and lively prose can compensate for a lack of clear purpose and efficient organization. And, certainly, no elaborate or beautiful printing and artwork can compensate for clumsy ideas and awkward language (although many reports seem to try this maneuver). But even the smallest editorial errors made at the last moment often can and do undermine several days of good thinking and clear writing. You must be vigilant until the very last moment of proofreading!

PLANNING

STEP 1.1: ASSESSING THE NEED TO WRITE

Sometimes, when I am feeling mischievous, I tease my clients and make them promise *not* to write at least one thing in the next six months.

The point I want to make is that there is too much writing. Typically, the people who write much of it wish they did not have to; the people who have to read it are rushing to sign up for "speed reading" courses.

Undoubtedly, most of us write more than we should—the exception being those few engineers and scientists who so hate to write that they must be forced with threats to commit a single sentence to paper. Most, though, write with little provocation, despising it most of the time, but writing anyway. Why?

Step 1.1 in the Writing System calls for you to assess carefully the need to write. Examine the impulse that made you think you should start a report.

There are three kinds of impulse to write:

- **Organizational** — any standard, conventional, periodic policy on what reports need to be done and when. Typically organizational impulses are tied to the calendar. (If this is the third Thursday, I must owe them a report.)
- **Psychic** — any emotional need to write, either to unburden strong feelings or

protect one's anatomy from danger, or, often, just to call attention to one's existence and prove that one is a good worker.

- **News** — an event, a transaction, a problem, or a situation that needs to be written about, or that needs writing to resolve a conflict or ease a pressure.

Most engineers and scientists feel these impulses all the time. Moreover, each impulse creates its own kind of stress that can only be relieved (they think) by writing something.

Whenever you feel one of these impulses, you must *decide* whether you will write. Actually, you have four alternative ways to respond to the impulse:

First, *ignore it* and hope it will go away. Some conventional organizational reports can be skipped without notice or harm. (Of course, I do not want you to risk your job. But do experiment to see whether your biweekly report would be missed.) Many psychic impulses, like anger, pass in a few hours, long before your memo would be back from the typist.

Second, *speak.* Speech can do almost anything that writing can—and a few things it cannot (such as calm an angry client or sell to a reluctant sponsor). Whenever your impulse can be satisfied by talking (on the phone or face to face), do it. Speech is actually a *better* way to satisfy most of your impulses; you should almost always prefer it.

Third, *use "boilerplate."* If you have no compelling reason to write, or nothing important to say, you can use old documents, "boilerplate": earlier reports, standard passages, form letters, company literature, or anything that has already been written and can be cut and pasted into acceptable form. (Of course, much cutting and pasting nowadays happens in an automated "word processor.")

Fourth, *write.* If you cannot ignore the impulse, if you cannot substitute speech, if you cannot contrive some boilerplate . . . then you need to write. And write at a level of quality suitable for the occasion.

I suspect that a careful assessment of the impulse to write will allow you to write less and, therefore, spend less time writing. At the lower levels in the organization, where writers have fewer choices and less power, this assessment may only save 5 or 10 percent of the time spent writing. But at the higher levels, where people have greater control over their own work and responsibilities, this assessment might reduce the time spent on writing by 20 to 40 percent.

(For those of you who do not write enough, I fear I have no savings.)

The rest of the Writing System is for those times when you decide to write and, therefore, to write as effectively as you can.

STEP 1.2: DEFINING THE SUBJECT AND OCCASION

The first, and easiest, part of the planning task is to define your subject and occasion.

I call it the easiest part because it comes so effortlessly to people with a technical or scientific background: people who are, in effect, *subject specialists.*

Paradoxically, defining the subject is usually the only strategic planning that writers do, and, taken alone, it can distort or harm the plan. Knowing the subject is useful and important in planning only if it is the *first part* of the plan.

The aim of Step 1.2 is to assign just a descriptive name to the product you are about to write, a string of words that you can use to talk about it, or create a file for it, or call a meeting on it. Do not try to put too much planning or deciding into the subject; keep to simple phrases:

- Woodbridge Water Treatment Proposal
- Semiannual Cost Analysis
- Risks of Chest X-Ray
- Five Candidate Warehouse Sites
- Factors Inhibiting Tumor Growth
- Counseling for Compulsive Gamblers
- Technical Supplement to the Manhattan Plan
- ABCO Affirmative Action Plan

Sometimes the report or letter you are about to write does not have a subject as much as an occasion, a set of circumstances that call for writing:

- Kansas City Trip Report
- Notes on ABCO Site Visit
- Reply to Jones's Invitation
- Follow-up on Smith Meeting
- Agenda for Staff Meeting
- Briefing Package for Presentation

Notice that in both cases, whether you use the subject or the occasion, you say very little about what the final product will look like or contain. *Saying little is the correct approach to defining the subject.* It would be a mistake to go further and write something like:

- Proof That X401 Inhibits Type-R Tumors
- Technical Supplement to the Manhattan Plan Justifying a Change in Discount Rates
- Report on the Kansas City Trip, Recommending That We Not Acquire the Elmwood Property

There will come a time later in the planning process for statements like these, but that time is not now.

STEP 1.3: ANALYZING YOUR FUNCTION AND PURPOSE

Now is the time to decide just what kind of product you are designing: to determine its function and purpose. This is the moment to think like an engineer or manager (not like an artist or composer) and decide what this written product is really *for.*

Do not be confused. The question is not the same as the one we raised a moment ago: not what it is *about,* but what it is *for.*

There are many classification schemes with which you can describe the several functions of business and professional communication. The approach I want you to use in Step 1.3 appears in Table 3-1.

Notice that there are just three broad functions:

- **Informing** — adding to what people know about matters of fact
- **Persuading** — changing people's attitudes (that is, what they like); opinions (that is, what they agree with); and beliefs (that is, what they consider true)
- **Motivating** — causing people to do something they might otherwise not have done

INFORMING (ADDING KNOWLEDGE)	• Telling or asking about facts • Describing things and events • Abstracting, reporting on systems, projects, and processes • Teaching procedures
BORDERLINE CASE	• Giving undisputed opinions
PERSUADING (CHANGING BELIEFS)	• Adding opinions about matters of fact • Superseding opinions about matters of fact • Changing preferences and priorities • Making people satisfied or dissatisfied • Justifying an action or assertion
BORDERLINE CASE	• Winning a promise to act
MOTIVATING (CAUSING ACTION)	• Making workers work • Making providers provide • Mobilizing people and resources • Altering policies or practices • Getting people to buy or spend

TABLE 3-1 Functions for Professional Communication

Now, as you read this you may suddenly remember something you heard once in a course in speech or psychology or "semantics"; you may remember being told that no one can really get someone to believe a statement or perform an action, that people can only convince *themselves* to believe something or do something. You may think that when you write all you do is inform, give facts and ideas, and let the readers reach their own conclusions or deduce their own actions.

There is a sense in which this view of things is correct: Ultimately no one can really make another person change his or her mind unless the person is willing. Indeed some people can hold on to their beliefs even through beatings and tortures!

But there is another sense in which this view is nonsense, akin to the notion that no one ever "really" understands or communicates with another person (a view held, remarkably, by many people who call themselves communication experts). Good writers know that effective writing *can* inform and persuade and motivate, and that all three happen with impressive regularity.

Of course, good writers also know that they cannot persuade and motivate —or even inform—without exploiting their understanding of the reader's perceptions and ideas. So, writing in itself cannot change a person's beliefs unless it has been written in a way that starts with the reader's perceptions as a "given" and moves those perceptions to the point where the writer wants them to go.

The functions of informing, persuading, and motivating are further divided, typically into smaller subfunctions.

As Table 3-1 shows, *informing* usually consists in telling about facts or describing objects or systems, reporting events and phenomena, or giving technical instructions. Informing is, by definition, neutral, dispassionate, objective, detached, and strictly descriptive and factual. The moment you tell your reader that a certain fact is important, you have stopped informing and started persuading.

Persuading consists mainly in changing the reader's opinions (or attitudes or beliefs) about what is true (whether a matter of fact for which there is little evidence or a matter of opinion), or about what is good or desirable. Many persuasive messages are also intended to make the reader feel either happy and satisfied or unhappy and dissatisfied.

A borderline case between informing and persuading is *giving undisputed opinions.* That is, there are times when you express an opinion or judgment (which would usually be part of a persuasive communication), but because you are a trusted expert with no one to challenge you, your opinion is treated as though it were a fact.

Motivating consists in getting people to do things (perform, provide, organize, spend money) or sometimes getting people to get other people to do things (changing the behavior of organizations and institutions).

The borderline case between persuading and motivating is *winning a promise to act,* a result that usually—but not necessarily—results in action.

The distinctions among these three broad functions are not always sharp and immutable. (I might quip, as one author has, that the existence of borderline cases proves that the borderlines are real and clear.) But I cannot overstress the importance of realizing that *there are three distinct functions of professional writing, and each function calls for a different approach.*

Put bluntly: Unless you understand which function you are performing, and know what kind of writing is required to perform that function, you will fail as a writer!

Writing to Inform

Writing to inform is the simplest of the functions—although doing it well can be hard enough to weary most writers.

An informative communication has a simple goal: to put some words and images into the mind of the reader that were not there before.

The easiest kinds of informing are those messages in which the readers *want the information,* where they have asked you a question, or where they are waiting for some result or milestone report. I call these the easiest messages to write because they will usually interest the reader, even if you have taken no special steps to make your work interesting.

In cases where you want to capture the readers' interest, however, to make them curious about a new body of facts, a new system, an event, then you are obliged to write better, to "hook" the readers into attending to your facts and information.

But wait. Why are you trying to interest the readers in some new information? Are you trying to change their minds? Win their favor? Sell them something? If the answers to these questions are yes, you may really be trying to persuade or motivate, not inform.

Some communication theorists argue that there are no cases of "pure" information. They claim that every aspect of your report—the order in which you present data, the way you lay out the pages, the tone and style of your sentences—all are intended to affect the interests and beliefs of the readers, that you are, in effect, persuading.

Probably these critics are right: There is no "pure" information. But so what! Every engineer, scientist, and technical professional has written scores of informative reports, pure or not. Their only aim was to move hard data and neutral, descriptive information from one mind to another, often only to comply with some periodic reporting requirement.

From a planning point of view, you should know that informative communication derives its entire impact from *clarity* and *accessibility.* If an informative message is clear (that is, resistant to misunderstanding) and accessible (that is, capable of being read and understood with reasonable effort), then it is effective. In that sense, informative communication is not easy, but the easiest kind.

Writing to Persuade

Facts do not persuade; data do not conclude; information does not shape opinions. To change what a reader believes, clarity is necessary but is not sufficient.

Getting readers to modify their beliefs—especially to revise long-held ideas—takes energetic, persuasive writing. The report must be not only clear but *cogent.*

Persuasive writing *builds a case.* The conclusion of the case is the idea or belief you want your reader to agree with or accept. The facts and data you introduce are the premises in the argument or the evidence in the case.

Later, I shall include a simple lesson on the ways to prove a conclusion. Now, though, the point to be stressed is that you must learn to recognize a persuasive message when you are about to write one. Is the purpose of your message to change your readers' minds? Make them like something more? Have a different opinion of you? Become more (or less) concerned about an issue?

If the answer to any of these questions is yes, you are about to persuade. You must understand that the message you are designing needs more parts than an informative message: It needs conclusions and judgments, and it needs logical bridges that connect the facts and information in such a way that they support or lead to that conclusion.

(Remember the borderline case: giving an undisputed opinion. In that instance there may be no need for evidence or logical support.)

If you are attempting to change minds on an important or sensitive issue, you may have the additional responsibility of helping your readers to agree with you in a way that reduces their own sense of inconsistency or disloyalty. Most readers will not change their minds about important things because to do so forces them to admit an earlier error or naiveté; if you cannot help them through this awkwardness, they will probably reject your new idea.

Further, if your persuasive message is one among several conflicting and competing messages, your document may also need to include refutations of your critics' and competitors' arguments. You may have to, in effect, write the winning case in a debate.

Persuasion is complicated enough to merit a life's study. Yet, despite the thousands of person-years of thought and research that have been spent on it, we know little more for certain about the subject than Aristotle knew when he wrote his handbook on persuasion, *Rhetoric.* (Even today, some of the people who study persuasion call themselves "rhetoricians.")

We know that at the time a person receives a persuasive message, he or she cares considerably about the reputation and prestige of the person who is doing the persuading. The reader will trust a trustworthy sender and mistrust an untrustworthy sender. Yet, as obvious as this relationship may seem, we also know that after a time people *forget* who sent them certain messages and begin to believe the ones they originally doubted.

We also know, much to our distress, that any assertion made often enough begins to win believers. (Just as professional grant writers know that any organization or person, no matter how wretched, can get at least one or two lavish testimonials.)

Being persuasive, therefore, requires you to do more than present the facts in an accessible way. It takes argumentative skill and a certain intellectual resourcefulness—a knack.

Writing To Motivate

Readers are moved by what they read, not just in an emotional sense but in a literal, physical sense. By writing to motivate I mean, literally, getting people to *do* things that they would probably not otherwise have done. (In some cases, the aim of motivational writing is to get people to keep doing what they are already up to; presumably, the writer believes that without this added motivation the readers would slow down or stop performing.)

To motivate readers you must do more than get them to agree with your conclusions: You must convince them that it is in their personal interest to do what you want them to do. Motivational writing *may* use persuasive arguments along the way to achieving its purpose, but its impact comes from showing the readers the advantages (benefits, gains, or inducements) they will receive by doing what you want them to do or, conversely, the disadvantages (costs, losses, or sanctions) they will experience if they fail to do what you are advocating.

When I talk to engineers and scientists about writing to motivate, many of them protest. They think that motivating (or even persuading and motivating) has nothing to do with the reports and papers and correspondence they write, that it is more properly the concern of advertisers or politicians or company executives. This is fallacious thinking.

Every report that contains a recommendation is supposed to motivate. Every proposal, bid, or quotation is supposed to motivate. Every paper that calls sincerely for further research is supposed to motivate. Even a simple memo reminding someone to bring certain data to the Monday staff meeting is supposed to motivate.

Motivational writing can have small, simple aims (like getting a particular item added to an agenda) or grand, complicated aims (like getting a government agency to finance a multimillion-dollar project). Regardless of its scale, however, it always uses motivational materials and ideas—what some people call *appeals.* Appeals are those things (or feelings or values) that people want more of, or want to avoid losing. To the extent that you can offer these appeals as inducements to your readers (more wealth, more power, better health, and a few others), or to the extent that you can threaten the loss of these valuables (higher costs, less independence, worse health, and so on), you can get readers to do what you want.

Of course, some people find this discussion of motivation uncomfortable; they consider this talk about appeals, inducements, and sanctions "devious and manipulative," not "open and honest." There are even engineers and scientists who consider it just plain wrong, "unprofessional," for them to tell people—customers, bosses, anybody—what they ought to do.

Such writers, however, are never likely to get anywhere as communicators (no matter how clear their sentences and paragraphs become). And they are not likely to go as far in their careers as those who know when and how to motivate.

Some Unclassified Functions

In my own work, the functions of *informing, persuading,* and *motivating* are broad enough and rich enough to explain nearly all my purposes for writing. There are, however, a few special cases that are harder to classify.

If there were a fourth category, it would have to be *entertaining,* writing to divert or amuse the reader. Surely, there are many articles and letters (even a few business memos) intended to make their readers smile. More usually, though, *entertainment* is a salutary effect of writing that has some other function; to say that a memo or report is entertaining is equivalent to saying that it is written in a lively, interesting way.

There are also many messages purportedly written to "express" the strong feelings of the writer: to register anger or protest, or "go on the record," to call attention to oneself. And, similarly, there are many messages intended to "impress" the reader with how smart, talented, eager, and worthwhile the writer is.

For myself, I try to decide whether this expressive or impressive impulse is really an impulse to communicate to someone else (that is, to inform, persuade, or motivate some readers) or just a form of ceremony or socializing. In other words, am I writing about my anger because I want someone to solve some problem (motivation), or just because I feel one of those psychic urges to get something off my chest? Am I expressing a political opinion because I want people to agree with me (persuasion), or because I am eager to enjoy the stimulation of political discourse? (Many people regard the expressing of opinions as a sacred right; they do not care who agrees with them, or even who listens. This kind of "communicating" is more like a religious ritual than an exchange of messages.)

Almost every message has at least some of this expressive or impressive function. (Every good message has a bit of entertainment as well.) The point to be stressed is that these are *incidental to the main functions* of your writing. No matter what personal or psychic impulses you have, you *must* decide what you want your readers to know (inform), believe (persuade), or do (motivate) as a direct result of your message. Until you decide that, you are not prepared to write.

STEP 1.4: ANALYZING YOUR AUDIENCE

While you are thinking about your function and purpose, you will inevitably be thinking about your readers or audience. Obviously, your aim is to inform, persuade, or motivate someone or some group.

Analyzing your audience can be a simple matter of thinking a minute or two about the person to whom you are writing. Or, in the case of proposals, it can entail several person-days of research and investigation.

Before I describe the details of the analysis, however, I want you to be aware of an important rule that most writers follow:

WHEN YOU WRITE, PICTURE YOUR READER.

If you know the person you are writing to, picture that person—while you are planning, and drafting, and revising. When you are in doubt about how to begin or what language to use, again picture that reader and judge what is appropriate for him or her.

And if you do *not* know the person you are writing to, or if you are writing to a large group, picture your intended reader anyway. If necessary, make up an imaginary reader (or two if there are two quite different kinds of readers) and keep that picture in mind.

(Throughout the planning and writing of this book I have been picturing two readers, composites drawn from the thousands of scientists and engineers who have taken my seminars. Repeatedly, I have revised my approach or thrown out pages because I knew that these two readers would find what I had just planned or written neither interesting nor useful.)

You do not need to be right about every aspect of your audience. What is important is that you always remember your audience and write to *them* instead of to yourself.

Picturing or visualizing your readers is a kind of informal, shorthand way of analyzing them. When you are planning (Step 1.4), you should also address some formal questions about your audience:

- What are the *basic,* unchanging (or slow-changing) characteristics of my readers?
- What characteristics of my readers are relevant to this *particular* message I am writing?
- What are the *communication preferences* of my readers?

Basic characteristics are the general, descriptive features of the readers. In communicating about technical subjects, the most important features are education and occupational experience. Nothing should be more obvious to a writer than the need to use language and concepts that are familiar or accessible to the reader. And yet nearly every day I see reports, newsletters, proposals, and even "nontechnical" introductions filled with references and technical vocabulary that the reader will not recognize. (I do not have to speculate about whether the language is unrecognizable; I am, in fact, the intended audience and I do not, in fact, know what they are talking about!)

Another important basic characteristic of the audience is *age.* Not because the age of the reader matters in itself, but because age is a kind of synthetic measure of many characteristics of the readers: when they were educated; what science and technology have appeared during their working lives; which social and political epochs shaped their values; how near (or far) they are from retirement.

All of these factors—compressed into the index of age—affect the interests and motivations of the readers, even the style of language that is acceptable. (For example, I find that many people educated during the Vietnam era, 1968–72, tend to consider proper grammar a kind of "establishment hangup." Talented engineers educated in that period are usually astonished when their bosses tell them they cannot write; no one told them that the whole time they were earning A's and B's in college.)

There are also several characteristics that can be classified as prejudices or loyalties, depending on how you look at it. Put simply, there are many beliefs and principles felt so strongly by your readers that they are deaf to any criticism or complaint about those beliefs. Indeed, they will resent or fear you for challenging or doubting those beliefs. It would be terribly hard, for example, for you to convince a conservationist from western Colorado that America ought to speed up production of synthetic oil. Or to convince hospital administrators that several hospitals in a community should share a single tomographic scanner. Or to convince a company that manufactures videodiscs in format A that the entire industry should be standardized to format B.

(In my own work, for example, I find it hard to convince executives that they cannot dictate a business letter without making a stylistic error, or to convince feminists that the pronouns *he, his,* and *him* should be allowed to apply to both sexes.)

Understanding these basic characteristics helps to define which purposes are attainable or unattainable; also, which "strategies" or "styles" will work best on this particular occasion. At the very least, considering these characteristics will tell you if you will have an easy or a difficult time in getting through to this reader.

Particular characteristics are those that bear on the current transaction, the message you are planning right now. The most important questions are:

- What is my relationship to the reader(s)? Are they superiors, customers, supporters, stockholders, suppliers?
- What relative power does my reader have over me? Who needs whose approval? Who is doing whom a favor? Who merits respect and deference? Who should sacrifice most? How far should I go to please?
- How does the reader feel about me? Am I trusted? Liked? On trial? Under suspicion? Revered?
- What notions or beliefs does the reader have about the subject of the current message? Is this new information? Does it conflict with loyalties or strong preferences? Will my message be in conflict with other messages?
- What exactly do I want the reader(s) to do as an immediate consequence of reading this message? (Notice how close this question is to the question of purpose; the two analyses always come together.)

You need not answer every question in detail. Rather, consider each question for a moment and decide whether it is important to your current plan. Typically, one or two of these questions will jump out, impressing you as being extremely relevant to your message.

The question of trust, for example, is essential. A motivational report written to a highly trusting reader, for example, need contain only recommendations; your trusting reader will probably accept them uncritically. The same report written to a skeptical audience will be five times as long and, even so, far less likely to achieve its purpose.

This analysis of particular characteristics is most relevant in the planning of proposals. (Ironically, proposal writers, particularly scientific researchers and consulting engineers, often ignore the characteristics of their audiences altogether and then wonder why they are not winning as many grants and contracts as they should.) Indeed, the winning margin in a successful competitive proposal (assuming the bidders are equally qualified) is how well the proposal appreciates the peculiar attitudes and interests of the sponsors.

In addition to these general and particular characteristics, most audiences also have *communication preferences,* favorite ways of receiving messages. Some audiences insist on an oral briefing before a written report. Some insist on a large number of pictures. Some like wide margins with few words to the page; others are annoyed by wasted paper.

Readers' preferences for various media or formats or styles can affect their perception of your message, often in ways the readers themselves do not understand. For example, most government agencies insist that they will penalize a contractor or offeror who submits "an overly elaborate" proposal. Ostensibly, a proposal that looks too extravagant or expensive will be regarded as evidence of wastefulness or irresponsibility. *In fact, most proposal readers are impressed favorably by high-quality paper, printing, and artwork and are put off by cheap or flimsy production techniques.* (In several years of working with proposal-writing engineers and scientists, I have *never* been told of a case in which an offeror was penalized or even scolded for an elaborately produced proposal.)

Think also about the "channels" through which you will send your message. Would it be better received with your boss's name on it? Should it be presented privately first, or to the whole group? Should it be long or short? Is it acceptable to send an unfinished draft?

Again, only a few of these channel questions will be relevant. And if you think about them for a minute, you will know which ones.

WRITING FOR A DOUBLE AUDIENCE

So far, the Writing System has encouraged you to think about *the* reader or *the* audience, as though there were only one. In fact, most of the longer messages written by engineers and scientists have *two* audiences or sets of readers.

What exactly do I mean, though, by *two* audiences? Are two readers automatically two audiences? (Are fifty readers, therefore, fifty audiences?)

Obviously, the number of readers does *not* equal the number of audiences: Simply, an *audience* is one or more readers *for whom you have a single purpose and a single set of audience analyses.* Thus, if you were writing a plan for a municipal planning group and its staff, in which your purpose was to get them to approve the construction of a waste treatment plant (one purpose for all of them), you would best consider them *one* audience. But if you had included in your report material that was highly technical and intended only for the group's two staff engineers, you would have had two audiences.

Longer proposals and reports typically have two audiences: one technically trained, the other administrative or managerial. Government-sponsored R&D reports are read by program scientists *and* contract administrators. Engineering reports in a company are read by engineering management *and* marketing or product development executives.

Again, a group becomes two audiences either when you have two purposes corresponding to two subgroups or when you take different approaches to language, knowledge, and other audience factors for two subgroups.

But is it possible to communicate to two different audiences in a single message? What if the purposes are radically different for each audience? What if the background and training of the audiences are miles apart?

Writing for a double audience is always difficult and problematical. The bigger the gaps between the audience, the harder the task. The simplest way to deal with the problem is to prepare two *separate* messages instead of one, or, what is more likely, to include a separate section or sections for one of the two audiences so that each audience in effect reads different parts of the same document. Summaries and introductions are examples of front materials included for the benefit of one of the two audiences; technical appendixes are an example of end materials included for the same reason.

If you have two audiences and do not have the option of writing separate messages in separate sections, however, then you must do a bit of added planning. First, decide which of the two is your primary audience. Because writing for two audiences forces you to bother one audience with material more suited to the other, you should think of the primary audience as *that reader or group you are least willing to inconvenience.* In other words, your primary audience should be less aware of the material intended for your secondary audience than the reverse.

Figure 3-1 shows the relationship between the primary audience and the secondary audience, a relationship that can assume three forms. (You have only *one* secondary audience, but it may be one of three kinds.)

For the "close secondary" audience, that is, readers very nearly as important to you as the primary reader, you may interrupt the flow of your text "in-line" so that, for example, a technical term, suitable for your primary reader, is followed immediately by a definition or explanation aimed at your secondary reader. For example:

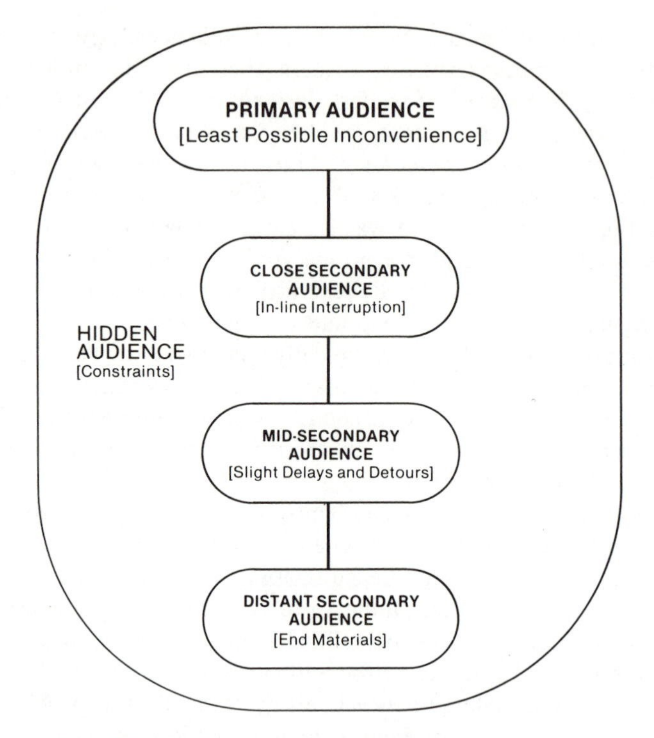

FIGURE 3-1 Relationships Between Audiences

1. **Single, technical audience:**
   The delay in the testing of the wing design has exhausted the secondary slack for that component but not affected the primary slack for the X501 Project.
2. **Primary audience technical; close secondary audience:**
   The delay in the testing of the wing design has exhausted the secondary slack for that component (that is, used up the time allowed for that phase) but has not affected the primary slack for the X501 Project (that is, without changing the likelihood that the X501 will be delivered on time).
3. **Primary audience nontechnical; close secondary audience:**
   Although the wing design was not finished until the latest allowable date, we expect no change in the overall completion date for the X501. (That is, exhausting the secondary slack for the wing design has not affected the total primary slack in the project development network.)

Version 1 is clear enough for an industrial or management engineer (obscure to almost everyone else). Version 2 is still clear enough for the technical reader, but the immediate, in-line explanation is in great danger of annoying that primary reader. Similarly, in Version 3, the immediate technical restatement may distract or irritate the nontechnical primary reader.

The only time you should plan to accommodate your secondary reader in this distracting way is when he or she is close in importance to the primary, so close that you can risk the inconvenience.

Incidentally, be aware that there is a tendency among technical people to automatically consider the technical reader primary; this prejudice excuses the author from worrying as much about the secondary audience. In practice, though, nontechnical readers are more likely to be primary than technical, especially when you are communicating to higher levels of the organization.

In most cases, the secondary audience is *mid-secondary,* considerably less important to you than your primary. In those cases, you accommodate the secondary reader with footnotes and other slight delays and detours. Consider this version:

4. **Primary audience technical; mid-secondary audience:**
   The delay in the testing for the wing design has exhausted the secondary slack for that component but has not affected the primary slack for the X501 Project.*

   **Slack* refers to the amount of delay that can occur without causing the project to miss its deadline. *Primary slack* refers to the entire project; *secondary slack* refers to some particular milestone or target before the end of the project.

In this case, the secondary reader is accommodated with footnotes, end-of-section notes, references to the next chapter, and so on. How long a detour or delay you should use depends, again, on how close a second the secondary audience really is. The longer the detour or delay, the less your primary reader will be aware of the material intended for the secondary reader.

Thus, in the final case, *the distant secondary audience,* you may refer your secondary reader to end materials: glossaries, appendixes, reference tables, and other devices that—with some difficulty—can help or accommodate the secondary reader. Of course, forcing a reader to refer continually to glossaries is a good way to tell this reader that you consider him or her unimportant. Use the technique judiciously.

In addition to the primary and secondary audience, Figure 3-1 also shows the "hidden" audience, which is, strictly speaking, not an audience at all.

The hidden audience comprises all those people who may eventually read the message you are writing, even though you have no wish for them to read it. Unlike a true audience—an individual or a group that you really want to reach with some purpose—the hidden audience is a constraint on you, a force that prevents you from writing certain things that would be better left unseen by the hidden audience.

If you are writing about controversial or sensitive issues—energy, the environment, defense, economics—your hidden audience may be the entire world, which would be a terrible inhibition for a writer.

(I tell some of my clients to write nothing that they would be embarrassed to see on the front page of the *New York Times;* I know this is a burden for them and that it takes some of the candor and liveliness out of their writing. But the hidden audience must be accommodated anyway.)

So far, then, I have talked about two real audiences, primary and secondary, and one quasi-audience, the hidden audience. Can there be three true audiences? Or four?

For practical reasons, no. To plan and write in a way that is appropriate for two very different audiences already taxes the resources of even good writers. Writing for three is almost like juggling three balls, except that there are far more people who can juggle three balls than can write effectively for three audiences.

If you are in a multiaudience dilemma, then you must abstract *one* or *two* audiences from the many. As I said before, form a composite that merges several individuals into one reader. Even if you are writing to a broad and ill-defined audience (such as one for a book or magazine article), your best course is to imagine one or two typical readers and visualize them in detail!

You are safer with one or two clearly defined imaginary or composite audiences than with three or more real ones. And you are never in greater danger than when you write to an undefined or "universal" audience.

STEP 1.5: DEFINING PRECISE OBJECTIVES

By now in your planning you have selected your purpose and function, you have analyzed your audience(s), and you are ready to define your precise communication objective.

This communication objective is the most important part of your plan; it "drives" the organization and outline, and it measures your effectiveness.

In operational terms, your *communication objective* is defined as

> *Precisely the circumstances under which you would conclude that your message was effective, that is, precisely what your reader should know, believe, or do as a direct and immediate result of reading your message.*

Consider these examples:

- The supervisor will know that all five tests have been completed (informing).
- The operator will understand how to log on (informing).
- The sponsor will accept that the project is on schedule, not behind (persuading).
- The CEO will have more confidence in this forecasting model (persuading).
- The company management will stop production of this brake model (motivating).
- The training department will offer health education for all employees (motivating).

The more precisely you can state what you want the audience to know, believe, or do, the more likely you are to write a message that will achieve that effect. The difficulties in defining the objective are many. Many writers *do not know exactly why they are writing,* especially if the document is in response to vague organizational pressures. And, finally, there is the problem of *cali-*

bration, knowing exactly where in the chain of communication and action to set the objective.

Figure 3-2 illustrates this problem. The criterion objective is your true objective, called "criterion" because it sets the standard for success or failure. That is, if your memo or letter or report does not produce the intended effect, it is a failure, regardless of how "well" it is written.

The first difficulty is in distinguishing between intermediate objectives and the criterion objective. For example, to get the company to stop production of that brake you must first persuade the management that the design is unsafe. But this persuasion objective is only a means to your real end: getting the audience to stop production. (If they agree that it is unsafe but do not stop production, your report is a failure.)

To distinguish an intermediate from a criterion objective you must ask this basic question: If Objective O occurs, and nothing else, will I consider the message a success? If the answer is yes, then O is a criterion objective. But if the answer is no (usually because something else must happen as well), O is your intermediate objective and that "something else" is your true objective.

The second difficulty is to distinguish criterion objectives from longer-range consequences. The issues are timing and feasibility. A criterion objective is achieved at once or within a short interval after the message is read by the appropriate audience. Further, the criterion objective is attainable, that is, feasible, given the time and resources in the situation.

You might decide, for example, that you cannot stop production with just a single document, that this is too difficult and complicated a goal for one communication. Instead, you might decide that your objective is "to get management to convene a meeting this week to discuss the possibility of stopping production." The actual stopping, then, would be a longer-range consequence of achieving your objective.

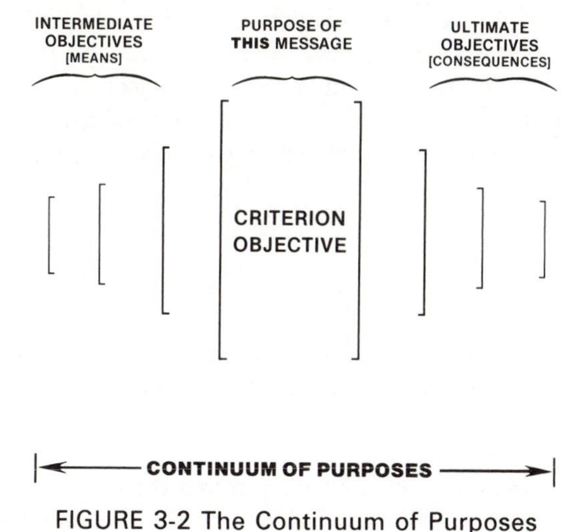

FIGURE 3-2 The Continuum of Purposes

When you have spotted the true criterion objective, you will usually recognize it: The language will be clear and simple; the effect will seem significant and relevant; the result will seem feasible and attainable.

STEP 1.6: DEFINING STRATEGIC PROBLEMS AND SOLUTIONS

You have now nearly finished the strategic planning of your message. (The rest of the planning is to produce the outline.) Step 1.6 is a kind of last-minute, catchall task in which you stop and ask yourself:

> What is likely to prevent me from achieving my purpose with this audience? What barriers, pitfalls, or obstacles are between me and my objective? What can I do to avoid them? Or even put them to my advantage?

Most barriers can be guessed or inferred from the characteristics of your audience. Barriers to informing are lack of knowledge and training, unfamiliarity with the material, lack of reading skills, lack of interest, and so forth. Barriers to persuasion are other and conflicting ideas, different from those you are advancing, including deep-seated principles or prejudices.

Motivation is the most strategically difficult objective. To move someone, literally, you must overcome the other forces that account for his or her current movement—including the powerful force of inertia that keeps everyone from wanting to change, and also the principle of least effort that keeps everyone from wanting to risk or sacrifice. You will need to think carefully about what forces are now pulling and pushing your reader, to find the best way to intervene.

Sometimes your analysis of barriers and obstacles will so discourage you that you abandon your objective altogether. (This is not always bad. You will save time, after all, by avoiding hopeless communication projects.) Other times, though, you may decide to revise or refine your objective.

At the end of this step, then, you should be confident that your objective is correct and that you have a way of averting any obstacles that will prevent you from reaching it.

DOESN'T THIS TAKE A LONG TIME TO DO?

It is much harder to describe the strategic-planning process than to do it. Except on the largest projects, most of the answers to the questions are straightforward and obvious. The strategic planning for a letter can take five or six minutes; for a three-page memo, ten or fifteen.

Whatever it takes, however, be assured that there is no more cost-effective way to use your writing time than to plan well. Five minutes of intelligent planning can save five days of revision; an hour of thoughtful planning can turn a proposal that loses into a proposal that wins tens of thousands of dollars.

Strategic planning is critically important. Writing without it is like trying to write without thinking; writing without a clearly stated objective almost never succeeds. And even when it does, no one is sure.

Organizing and Outlining

STEP 1.7: SELECTING THE FORM AND SCALE

Now that you have defined your subject, audience, and precise communication objective, now that you have appraised the communication preferences of your reader, now that you have predicted what obstacles or pitfalls lie between you and your audience . . . now you are ready to decide the overall shape and appearance of the message you are about to compose.

(Once in a while, I hope, you will have decided *not* to write but to telephone or visit your "readers" and talk to them.)

Your first questions, therefore, are the following.

What vehicle shall I use? A letter? A memo? A stand-alone report with a covering memo of transmittal?

How long, approximately, will the document be? Sometimes you can infer from what needs to be covered how many words or pages you will use. Other times you may have decided that it is good strategy to make the document especially short or especially long (for effect). And, in some rare cases, you may be writing a proposal or report that has officially been limited to a certain number of pages or words (or even a certain number of characters in some military communications).

Will the document have headlines or headings? That is, will there be clearly marked sections, with titles or captions or numbers to separate them? If so, will there be lots of headings (two to a page, for example) or few (one every five or six pages)? Will there be only one or two "levels" in the headings, or will there be an elaborate scheme with four or five levels of subordination?

Is there a standard format to be followed? Given the occasion, is there some standard report or document format that your organization wants you to use? How well will it fit your purposes? (How strictly enforced are the standards?) If it is inappropriate for your needs, can you escape using it? Can you modify it for your needs?

There are no straightforward rules to help you answer all these questions. The easiest technique is, again, to visualize your readers, but this time you picture them *actually reading your document.* How does it look in your imagination? What appearance and format seem best? And, even if you cannot picture the best or most appropriate format, can you at least picture what would be wrong or inappropriate?

Most professional writers have preferences regarding these issues of form and scale. Most prefer stand-alone documents if the message is to be longer than a page; a three-page letter should be rare, considerably rarer than three-page reports with two-sentence covering letters.

Most writers (and readers) prefer brevity—unless the brevity appears terse and unresponsive. (Some writers are afraid to be brief in their letters because a letter with only one or two short sentences looks so bare and forbidding; I advise these writers to get smaller stationery!) On those occasions when you suspect that you have to be long to be effective, be sure your suspicions are well founded. As long as your work is complete and sufficient (and sometimes that means *long*), you will rarely need to swell your document with useless and extraneous material. Although a long love letter is often taken as greater proof of love than a short one, no such relationship holds in business or professional communications.

(The controversial case is the *final report.* Most people believe that a final report should be long enough to justify the cost of the study or project. They do not want to turn in ten pages of report after $100,000 worth of work. In the mid-1960s, when dollars were worth more, I once heard a lab manager in an R&D company assert that final reports should contain at least one page for each $1,000 of contract—but not more than one and one-half pages.)

You must decide, then, the appropriate "heft" of the product, guided by the expectations in your industry or organization.

Most editors and writers also prefer that you use lots of headings and, therefore, lots of short sections—because that is the way most people like to read. Some companies permit headings even in short letters (and I, for one, approve of the idea). If your document will be difficult for your audience, use as many headings and sections as you can—and use long descriptive headings (a topic discussed later in this chapter) rather than short, uncommunicative headings.

Further, if there is a standard document format that you can (or must) use, by all means use it. Familiar structure usually helps the reader and the writer to follow the chain of thought. It is only when the standard format inhibits you from achieving your objective, from pursuing your strategy, that you should abandon it or resist it.

Consider, for example, such periodic reports as monthly progress reports, semiannual budget requests, and annual operational plans. Sometimes these standard instruments (with their standard formats) are the best way to capture your readers' attention. But there are other times when, for example, you want to put more urgency or surprise in your message, when you want to "interrupt regularly scheduled programming" to get extra attention. In those cases, putting your message into a standard format might diminish its impact.

(Incidentally, it is in the nature of all standard formats that they fit *no* case perfectly. Using them well, therefore, always entails adapting them somewhat to your needs. You must always decide whether it would make more sense to adapt the standard format or abandon it. The fact that the standard does not fit your needs exactly is a given.)

You will have completed this step in the plan when you can picture, at least vaguely, the actual physical appearance of the document: its look and shape and heft and parts. When you have a somewhat clearer idea of what sections it will contain, and in what order, you are ready for the next step.

STEP 1.8: ORGANIZING AND ASSIGNING THE "RAW MATERIAL"

Picture the last time you wrote anything longer than two pages. Picture yourself and your environment while you were writing—especially when you just began to write. What was on your desk at the time? Can you remember?

Most documents written in technical organizations contain items of technical information: statistics, specifications, diagrams, citations. And most of the time we write we view one of our aims (although certainly not our communication objective) as covering "all this raw stuff on my desk."

What kinds of "raw stuff" you have on your desk depends on your work. For example, you might have 150 wind tunnel test results, or eleven failed valves, or eighty-five complaining letters from clients, or twenty-four drawings of system subassemblies, or eleven case studies, or five proposals, or seven trouble-shooting reports, or 180 pages of cross-tabulations, or seven specimens of artificial lemon flavor, or five drawings for packaging artwork, or four letters of criticism from "peer reviewers." The list is endless.

For many of my clients, this pile of "stuff" is an obstacle. Not being able to "deal with it" (notice that the problem is discussed in psychological terms), they are slow to begin their writing. Until they feel "on top of it," they feel unready to write.

Step 1.8, therefore, is to appraise this raw material on your desk and assign it to some place in your document (or, in some cases, to decide to leave it out of your document altogether).

The first task is to *list* everything you have to "deal with"; giving things names usually makes them less forbidding. But, while you are naming and listing, you must group and cluster the items so that there are *no more than nine items or clusters of items in your list.* (I would prefer that there be no more than seven, but that limit might be too confining.)

The list should be short (nine or less) so that it will be manageable. Lists longer than nine (longer than seven, really) are too long to handle. You cannot see the relationships among the items; you cannot remember all the items at once. Your aim should be to encapsulate, in one manageable list, the undisciplined material all over your desk.

How are these clusters formed? You may group the items any way you like: by the calendar (months, periods, etc.); by geography (eastern states, southern states, etc.); by plant or location; by problem; by function; by client; by any classification scheme that might be used to organize or categorize the items.

Further, unlike an accountant, you need not be consistent, using only one accounting principle for all the items. The first three categories could be grouped by month, the last four by function, as in:

- July results
- August results
- September results
- Test plans
- Pretest protocols
- Sample lists
- Observation locations

Once the list is devised, you review each item (or cluster) to decide how you will dispose of it, that is, where you will place it in the document you are planning.

The main criterion, of course, is *relevance to the objective.* Items that have no bearing on your purpose can be discarded or, what is almost the same, relegated to the attachments and backup material. Items that bear directly on your objective can be featured prominently—at the very beginning usually.

(In some cases, an item may have no obvious bearing on your objective but is important to your strategy—your method for overcoming barriers and resistance. In that case, you might also choose to highlight the item.)

If you are planning a message that has sections with headings, you may use those headings to assign the items. If your document will have no formal sections, however, you may use rough assignments ("middle," "toward the end") to map the items. For example:

- July results—featured in Introduction
- August results—featured in Introduction
- September results—summarized in Appendix
- Test plans—Methods section

- Pretest protocols—save for the final report
- Sample lists—confidential; do not include
- Observation locations—include at end of Methods section

Notice that the effect of this simple listing and assignment is not only to relieve you of anxiety about the raw material to be covered but also to begin the creation of the outline. Once the assignments are made, the shape of the document will grow a bit clearer, giving you something solid to start with as you write the formal outline.

STEP 1.9: ASSESSING THE MAGNITUDE OF THE JOB

By now, even though you still have no outline, you have a clear enough sense of the emerging document to answer the next planning question:

Can I write this all alone? You must answer this question before you write your outline because the type of outline you will need depends, more than anything else, on how many people will be working on the document.

In many cases, of course, the answer is obvious. A one- or two-page letter is clearly something you will do solo. A two-hundred-page proposal, due the day after tomorrow, will obviously need seven or eight writers (not to mention five typists, three artists, and one and one-half editors).

Step 1.9, deciding the magnitude and scope of the writing job, is really a task of project management. To answer the question correctly (except in cases where the project is so small that there are few alternative ways to do it) requires you to think about the approximate size of the job, the time you have, the resources (money and talent) that might be brought in, the other work you are neglecting by *not* bringing in other resources, and so forth.

Any question a project manager might ask about the job he or she is beginning—personnel, schedule, costs, standards of quality—can be asked about the writing project before you. And, as in any project analysis, the hard part is making trade-offs. Can I save some time or money if I cut quality a little? Can I get the job done faster if I use more expensive personnel? Can I make good use of existing slack, that is, of people who right now need work to do, even if they are not the best people for the job?

Even the basic question in Step 1.9—Can I write this all alone?—involves trade-offs. Now that you have analyzed the subject, audience, and purpose so carefully, could you just turn the piece over to a subordinate and let him or her carry out your design? Or now that you have decided that certain items of information to be covered are going to be given a low-profile treatment in the back of the report, could you turn over some of those less-sensitive tasks to your assistant or to junior staff?

Would you be able to meet a difficult deadline better by bringing in more people? Or would the time spent in coordinating and managing a team of writers take up more resources than it would save?

If you have decided that two or more writers will be working on substantive and important parts of the document, then you have also committed yourself to becoming a manager as well as a writer—the person responsible for seeing that all writers understand their part of the whole, that the writing assignments are carried out, and, hardest of all, that there is some unity or consistency of tone and approach across all the pieces written by all the writers.

At the very least, if you have decided that this is a project for several authors, then you must write—or manage the writing of—an elaborate and detailed outline. The more writers, the more detailed. Just as it is always more efficient to revise a document in the planning and outlining stages than in the drafting and redrafting stages, it is always more efficient to have an outline that is complete, workable, and understood by every member of the team of writers before beginning to write the drafts. Indeed, in some cases you would do best to bring in the writing team (or at least the leaders, for larger teams) to collaborate on the more detailed parts of the outline.

If you are going to write alone, you will usually need very little outline—just enough to keep you on track. The exception is when you are writing a document that must be approved by a supervisor, in which case you will want your outline to be detailed enough for your supervisor to appraise, so that you can both make changes in the outline *before* the draft is written.

Authors working alone usually need few notes—especially when they know their objectives. I know writers who can do a twenty-five-hundred-word technical article with just twelve words of outline, and a one-hundred-page monograph with only two pages of notes. Once you know where you are going (your objective), you also know where to start; once you have appraised and assigned your raw material, you are not likely to be sidetracked.

Whether your outline is simple or detailed, however, you should know something about the ways outlines work.

THE "THEORY" OF OUTLINING

An outline is a two-dimensional matrix that is forced upon a multidimensional array of information so as to produce a one-dimensional presentation. Is that confusing?

Outlining is a way to manage disorder, to bring wild and uncontrollable elements into a proper organization.

The key to writing an effective outline is to remember that the report or paper you are writing is going to be read in normal sequence, starting with the first line and ending with the last. Even if very few longer documents are actually read this way (people do skip and jump, forward and backward), nevertheless, a well-outlined message moves forward in a purposeful sequence. Thus, no matter what complex logical scheme you devise to organize your material, no matter how intricate your matrix or "conceptual framework," the final product is to be read linearly, one word after another, hay foot straw foot, through to the end of the piece. An outline is supposed to tell you what order to follow.

As I said above, though, an outline is a two-dimensional design leading to a one-dimensional presentation. Once you write your first outline entry, or heading, you can move on either of two axes—even though your document is moving forward in only one direction.

Consider Figure 4-1. In this representation, the direction in which your document is moving is *southward,* from north to south, the same way we read a page. Notice, though, that each time your message moves forward you, the author, have the option of moving eastward, rather than southward. Sometimes this eastward movement is shown typographically, with an eastward indentation:

I.
  A.
    1.
      a. (and so forth)

At other times, this eastward movement is represented with an eastward array of decimal partitions:

1.
1.1
1.1.1
1.1.1.1

By using letters and numbers this way, you can alert your reader to the eastward drift of your development, even while your text is still moving southward.

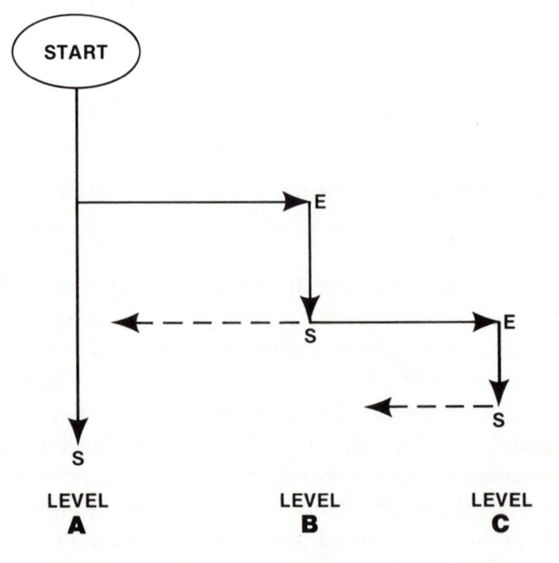

FIGURE 4-1 The Dimensions of an Outline

In many business communications, though, this eastward shift is shown without the use of letters or numbers, but with centering and typography:

FIRST HEADING

SECOND HEADING
Third Heading
    Fourth Heading

Each of these indentations or eastward shifts is called a layer or, more commonly, a *level.* (In Figure 4-1, the three levels are A, B, and C.) An outline need have only one level, in which case all the movement is southward.

Typically, though, an outline will have two or three levels (in shorter, one-person writing projects), or four or five in more complicated cases.

There is only one technical rule for outlines: There must be at least two entries on any one level before you move west from that level. Or, in more familiar terms, where there is an A there must be a B, and where there is a I there must be a II.

The following outline, for example, is flawed:

1.	I
1.1	A.
1.1.1	1.
1.1.2	2.
2.0	II

At the second level, there is a 1.1 but no 1.2, a IA but no IB. The outline moves west again before the second entry at that level, before the "second shoe drops," leaving the outline flawed.

The reason for this single rule is simple: Each eastward movement is a subdivision or partition of the material on the previous level; nothing can be partitioned or subdivided into only one part! Therefore, there must be at least *two* entries before you move west to a higher level.

(Note, even this one rule does not apply in every case. You should definitely follow it in your formal reports and proposals, but you need not follow it in your less-formal correspondence or memos. I recommend that you observe the rule in any communication where logic counts.)

All this talk of levels and directions is quite abstract. Even people who know how to write correct outlines might have trouble following it. But that is the point: Most people write outlines without being aware of the power of the tool, without appreciating how useful the craft of outlining can be.

There are only two logical moves in constructing an outline (east or south), and there are two sets of corresponding logical and communicational criteria to follow in preparing the outline: southward development and eastward development.

Southward Development

On any given level, there must be a purposeful, discernible sequence for the entries, from the first entry to the last, before a westward shift (return to a "higher" level).

Every time you construct a series of major sections or headings (at the first level), you must have some logic or sequence in mind. There must be some reason that Conclusions comes after Method (or the reverse); there must be some reason for the order of the major parts, some logic, and it should be clear to the reader.

The order of major parts at the first level is usually not a serious problem. In most formal documents the sequence of the main sections is determined by company standards or by long tradition. The bigger question is what to do about the lower levels.

Again the question is, Is there a purposeful, discernible sequence, and can the reader see it? Most correct southward sequences fall into one of these basic patterns:

- **Chronology** — the time sequence of the events discussed
- **Interest** — the degree of interest or fascination for the readers (as judged by the author)
- **Priority** — the order of importance or urgency (as asserted by the author)
- **Logic** — problem-solution, cause-effect, need-action, opportunity-plan, or other scheme
- **Difficulty** — easiest to hardest or hardest to easiest, complexity
- **Numerical sequence** — for numbered or coded entities
- **Alphabetical sequence** — for lists of undifferentiated items
- **Quality indicator** — order of desirability, height, weight, speed, price, size, novelty, originality, feasibility, availability, compatibility, or other kind of perceived quality or utility

Each series of entries on a given level (that is, until you move west again) must follow one of these organizing principles (or some other principle you derive). There must be some perceptible order of any kind; ideally, it should be a compelling and psychologically correct order to capture and hold the attention of the reader.

Test your outline. Even if the sequence of major sections is prescribed by a standard format, investigate the second level. What explains the order of items in your Introduction? Why are the three experiments in your Methods section in that particular order? Is there any logic to the sequence of tables in your Conclusions? What explains the arrangement of your five Recommendations?

At the very least, you must be able to give some coherent answer or other to these questions. If not, your outline has a flawed southward development.

Eastward Development

Before you make any southward movement, you should examine your latest entry to see if it needs subordination or partition. Recall that, strictly speaking, an outline needs only one level. Thus, the question of whether you will move eastward at any point, adding another level, is a question of craft and art.

Most of the criteria affecting eastward development derive from the *management* of the document, rather than its logic. Each time you move eastward you produce two effects: (1) you make the outline easier to write from, by providing more detail about the contents, and (2) you make the sections—the text between the headings—shorter.

There are, of course, people whose outlines keep moving eastward until they have the first sentence of each paragraph, so that the entries in the outline no longer correspond to headings but to all the paragraphs between headings. In my own experience, this ultrarefined outlining is appropriate only when you are teaching beginners to write; it is impractical and inappropriate for most professional writing jobs.

(In my own planning I like to have headings about every two pages. I try to write an outline that will have an entry every two pages or so—all the detail I need to guide me. Thus, before I move south in an outline, I estimate the number of pages it will take to cover the last entry; if it is more than two pages [or whatever other threshold I have set], I shall continue to move east, partitioning it smaller and smaller.)

A fascinating and peculiar aspect of outlining is that *the more you partition a section, the longer it gets.* There is no logical reason why this should be so, but it is a fact. In almost every case, if you partition a section of the outline into two or three subsections, the total, aggregate text under the original section heading will be longer than if you had just moved south instead of east.

STEP 1.10: PREPARING THE ONE-PERSON OUTLINE

With the possible exception of the one- or two-sentence cover memo or letter, every writing project needs a written plan and written outline. That is what I said: even one-page memos, even polite thank-you notes. For short and simple messages, the plan and outline will be short and simple—but there will still be a plan.

(If I did not insist that you write a plan and outline for every project, you might ignore the questions of audience, objective, and sequence for these smaller communications—many of which might be critical to your career.)

Most outlines begin with a rather straightforward sequence of major parts at the first level. In fact, I have found that almost every successful message has an outline that begins with this general framework:

Part 0:	Attention Getter
Part I:	Beginning of the "Body"
Part II:	Middle of the "Body"
Part III:	Ending of the "Body"
Part X:	Attachments

Of course, I am not suggesting that you use headings like these for your message (that is, assuming you're going to have headings). Rather, I am saying that your first outlining task should be to name, at the first level in the outline, the materials or sections corresponding to those five headings:

Part 0:	Title Page, Contents, Abstract, Summary
Part I:	Introduction, Problem, Background, Goals, Mission, Objectives, Purpose, etc.
Part II:	Approach, Method, Materials, Management Plan, Schedule, Design, Techniques, Scope of Work, etc.
Part III:	Data, Findings, Conclusions, Costs, Recommendations, Implications, etc.
Part X:	Attachments, Appendixes, Exhibits, Supplements, etc.

If you discover that you want two or more items *within* one of the five sections listed above, you have a choice: Will you find a single, first-level title that encompasses all the sections? or Will you have more than these five main sections at the first level? For example, suppose that in Part II you intended to write about Methods, Scope, and Schedule; you must decide whether you will have one heading at the first level—Technical Plan, say—that embraces the other three sections, or whether you want Methods, Scope, and Schedule as first-level entries.

In the following example, author A chose to keep to the five main sections, but Author B did not:

AUTHOR A	AUTHOR B
Summary	Summary
Introduction	Problem
Problem	Background
Background	Methods
Approach	Schedule
Methods	Results
Schedule	Conclusions
Results	Tables for Study #1
Data	Tables for Study #2
Conclusions	
Attachments	
Tables for Study #1	
Tables for Study #2	

Which outline is a better starting place? Would they not produce the *same* one-dimensional presentation?

My own preference is for Author A's approach. Notice that in A, the longest southward pattern is on the first level—five items. In B, there is only one sequence on one level, and it is *nine* items long. Nine items in a sequence is about as long a concatenation of information as most readers can grasp without help or explanation from the author—although Author B's pattern is sufficiently familiar that most readers can understand it. Try this less-familiar sequence, however:

Background
Economic Analysis Deficiencies
Possible Conflict of Interest
Level of Commitments
Supplemental Economic Analysis
Measures to Eliminate Possible Conflict of Interest
Changes in Levels of Commitment
CV (Curriculum Vitae) for Economist
Personnel Allocation Tables

For most readers, this sequence is a puzzle that can be solved with better-defined major sections:

Background
Problems and Deficiencies
   Economic Analysis
   Conflict of Interest
   Level of Commitments

Procedures to Correct Problems and Deficiencies
  Supplemental Economic Analysis
  Measures to Eliminate Possible Conflict of Interest
  Changes in Levels of Commitment
Attachments
  CV for Economist
  Personnel Allocation Tables

By turning the original first-level headings (a string of nine) into second-level headings, I change the outline so that the longest string is four and most are only two or three.

Nor is this the only way to revise the first outline. Consider the following:

Background
Economic Analysis
  Deficiencies
  Supplemental Analysis to Remove Deficiencies
  CV for Economist
Conflict of Interest
  Possible Conflict of Interest
  Measures to Remove Conflict of Interest
Levels of Commitment
  Insufficient Levels for Key Personnel
  Proposed Changes in Levels of Commitment
  Personnel Allocation Tables

There is, of course, nothing inherently wrong with a long southward sequence. If you have to describe 150 tasks in a schedule, you may certainly list the 150 as a series of entries in your outline—but it would help your reader if you clustered them into groups of 8 or 10, each cluster with its own identifier or name (even Phase 1 and Phase 2 if you cannot think up more descriptive names).

Notice also that in the specimen outlines you have seen so far there are two kinds of entries: categorical and topical. Categorical entries (Introduction, Approach, Method, Phase 1, etc.) merely tell you the *kind* of information to be included in the section, but not the actual material to be discussed. Topical entries, however (Economic Analysis Deficiencies, Conflict of Interests, etc.) give more insight into the actual discussion or text. Usually, categorical entries are suitable only for the first level in the outline, or possibly the first two levels; no outline, though, no matter how short, is complete until there are some topical entries. Indeed, the best outlines—as I shall show later—contain long and provocative topical headings (often called "thematic").

How can you know when you have enough outline?

If you are working all alone, if you are comfortable with your purpose and subject, a two-level outline with only twenty or thirty words may be enough. But if you are uncomfortable or uncertain about the subject, more detail, three or four levels, may be necessary. One entry per paragraph is usually too much detail for most outlines, except that there is nothing strange about having three entries for a three-paragraph letter or memo.

STEP 1.11: SECURING APPROVAL OF THE PLAN

If you have done all of your strategic planning—subject, audience, objective, strategic problems and solutions—you will need *less* outline than you would if your outline were your only plan. Your earlier strategic planning gets you ready to write, starts to form the words and pictures you need, so that you do not have to rely as heavily as most authors do on their outlines.

Ultimately, only *you* can know when your outline is detailed enough—unless, of course, the piece that you are writing is going to have to be approved or signed off by someone else (Step 1.11). Remember that the Writing System requires you to *secure approval of your plan and outline before you start your first draft.* Consequently, you will have to put in enough detail to communicate clearly to your supervisor what the final product will look like.

(And you must resist the temptation to conceal important features of the piece you are writing; those parts you are least comfortable about developing may be the very features that will surprise or distress your superior when he or she reads the first draft. To leave them out of the plan would be self-defeating.)

The aim is simple. Your plan and outline should be clear and complete enough so that your supervisor or boss can make an intelligent estimate of what the final product will look like. This person should be able to spot what he or she does *not* like before you have wasted the time and energy to write it.

You want to improve your message in the planning stage. Try several realignments of the parts and sections while it is still relatively easy to shift and relocate.

If you are having problems with the plan, get some help, either from your boss or from a colleague, or an editor if your firm has one. (Editors resent being brought in at the very last moment almost as much as writers resent having them brought in. If you involve the editor early, both of you will benefit.)

If may take four or five levels of outline before your supervisor understands what you intend to write, that is, before the supervisor finds something he or she likes or dislikes. But, even so, the effort is worth it.

Not only will you reduce the likelihood of writing something that needs substantial revision (caused, first, by writing without a plan and, second, by writing without a plan approved by your supervisor), you will also have made the first draft considerably easier to write. Some authors believe that if you keep writing more and more detailed outlines, the first draft comes out all by itself! (I am not sure, though. I fear Zeno's paradox might keep you from reaching the first draft itself.)

If in Step 1.9 (assessing the magnitude of the project) you decided that two or more authors need to work on this message—with you managing the process—your outlining needs will be much greater.

STEP 1.12: PREPARING THE DETAILED, "MODULAR" OUTLINE

Suppose that, rather than working all alone, you are now in charge of the preparation of a proposal. Here is your first outline:

Introduction
  Background of the Agency's Information Needs
  Current Information Projects
  Proposed New System
Approach
  Preliminary Design
  General Design
  Detailed Design
  Prototype Testing
  System Refinement
  Documentation
Corporate Capabilities
  Staff
  Company Experience

As outlines go, this is not a bad one; it has two levels, and the southward series at each level is intelligible. And it describes what is almost certainly the proper sequence of major sections.

As manager of this project, however, you should be able to answer certain questions about the proposal. Looking at the outline, can you tell me

- How long each of the sections will be?
- How long the finished document will be?
- How much artwork will be needed?

Or how about some more-substantive questions:

- Where will your company's main selling theme appear?
- Where will you describe the need to use Multiple Virtual Storage?
- Which staff will you describe?
- Which project summaries will you include?

Now put yourself in the position of one of the writers on the team, the one in charge of the "documentation" section. Looking at the outline, how will you know

- How much material to write?
- What sales pitches to make?
- How the documentation section ties in with the other sections of the proposal?
- How much existing material—"boilerplate"—can be used?

To continue this exercise, assume that you are the sponsor who reviews this proposal; now this outline is the table of contents. How will you know

- If the proposal addresses all of your specifications?
- Where each specification is addressed?
- Where you address the most critical issues?
- Where you prove your track record?
- Where you ensure that this is a turnkey project?

Do you get the point? This outline, even though it might be clear and complete enough for one writer working alone on a small document in a field he or she knows well, is inadequate for larger and more complex projects. It will not meet the needs of the project manager, the writers, or the reader.

The easiest way to correct this shortcoming is just to *add topical detail,* to move eastward for two or three additional levels of refinement, so that the words used to describe the sections give a clearer picture of what is to be covered in each small section. (As I said, some publication managers will outline right down to the topic sentence for each paragraph!) Adding this detail will solve many of the problems. But not all.

Adding detail will not tell you how long the sections will be; it will not even tell you which items in the outline are just headings and which require text. Adding detail will not tell you how much to write or, more importantly, what to *prove* or *sell* in the sections you are writing.

Nor will adding detail tell you how much artwork and graphics you will need, certainly not enough to help you to plan this most expensive and time-consuming part of preparing proposals and reports.

To make your detailed outline more useful, indeed to turn it into a sophisticated planning and management tool, you need an outline that adds three new features:

- First, an outline that uses more "thematic" and persuasive headings instead of the usual uncommunicative "noun strings" (categorical or topical);
- Second, a standardization of the parts of the document, so that your outline can address each standard part; and

- Third, a way of integrating graphics into the planning process, a way to marry text and pictures.

The first improvement, changing the headings from dull and descriptive to thematic and evocative, is a basic journalistic technique. Actually, it involves turning *headings* into *headlines*. The typical technical heading is a string of nouns with a modifier or two thrown in. Consider this four-level outline from a technical report:

2. Renovation Projects
   2.1 Acoustic Innovations
      2.1.1 Loudspeaker System Improvements
         2.1.1.1 Echo Problems in Outdoor Speaker Systems

Each of these headings is increasingly more detailed; each delimits the subject, makes it smaller and smaller. But none of these headings tells the manager or the writer or the reader what exactly is going to be covered or handled or explained or proved or refuted or criticized . . . or anything. Consider how different it would be, however, if for item 2.1.1.1 you had used one of the following headlines:

- THREE WAYS TO ELIMINATE ECHO PROBLEMS IN OUTDOOR SPEAKER SYSTEMS
- THE IMPOSSIBILITY OF PREVENTING ECHO PROBLEMS IN OUTDOOR SPEAKER SYSTEMS
- HOW ABCO'S NEWEST PHASE-DIRECTOR ELIMINATES ECHOES IN OUTDOOR SPEAKER SYSTEMS
- MISCONCEPTIONS ABOUT THE NEED TO ELIMINATE ECHOES IN OUTDOOR SPEAKER SYSTEMS
- THE NECESSITY FOR CONTROLLED ECHO TO PRODUCE PLEASING MUSIC QUALITY IN OUTDOOR SPEAKER SYSTEMS

These *headlines* (instead of headings) make clear what the theme and scope of the piece will be—not just to the project manager but to the writers and readers as well. They limit, guide, highlight, and justify the sections they introduce; they even provide previews of what is to be explained so that the reader can anticipate and follow what is coming.

Again, these headlines are *not* just a matter of greater detail; they are different in kind. If you were to assign one of your acoustical engineers to write a section on "echo problems in outdoor loudspeakers" and ask this engineer to submit a more detailed outline of the section before writing it, you would probably get still more neutral and descriptive headings, such as

- Direct/delay echoes
- Reflection echoes
- Outdoor terrain anomaly-induced echoes

Notice that with these noun and adjective headings the subject gets smaller, but the purpose and scope do not become much clearer. **This tendency to use neutral, descriptive headings probably accounts for the inability of many engineers and scientists to write persuasively; they might be better able to prove the point if they knew what point they were supposed to prove.**

Headlines (unlike headings) motivate the writer and the reader. They even prevent serious misreadings. If, for example, you were to read a section of a report on nuclear power plants and the section heading were *Coolant Pump Power Source Redundancy* (five nouns, no grammar), you might think that the section was about some inefficiency or excess in the design of the plant. Suppose, instead, the headline read:

- THE NEED TO ENSURE REDUNDANCY IN POWER SUPPLIES TO THE COOLANT PUMP, *or*
- PROVIDING ENOUGH POWER REDUNDANCY TO PERMIT SERVICING WITHOUT A SHUTDOWN, *or*
- BACKUP SYSTEMS TO BACK UP THE BACKUP POWER SOURCE FOR THE COOLANT PUMPS.

Now you get the message. The more redundancy in power to the coolant pumps, the safer the plant!

There is no single simple technique for converting traditional headings into headlines. But there are a few common forms:

- *HOW TO* (HOW TO SWITCH TO DISTRIBUTED PROCESSING)
- *NUMBERS* (THE FIRST THREE STEPS IN SWITCHING TO DISTRIBUTED PROCESSING)
- *ADVANTAGES* (HOW DISTRIBUTED PROCESSING REDUCES ADMINISTRATIVE COSTS)
- *VERBAL* (CONVERTING YOUR CENTRALIZED SYSTEM TO A NETWORK OF SMALL SYSTEMS)
- *INJUNCTION* (THE NEED FOR BETTER SECURITY SUBSYSTEMS IN DISTRIBUTED PROCESSING)
- *SALES THEME* (THE FIRST DATA PROCESSING SYSTEM THAT EVERY BUSINESS CAN AFFORD)

There are even some who write whole sentences as headlines:

- *DECLARATIVE* (DISTRIBUTED PROCESSING REDUCES COSTS)

- *IMPERATIVE* (CONVERT TO DISTRIBUTED PROCESSING BEFORE YOUR COMPETITORS DO)
- *INTERROGATIVE* (CAN THIS DATA PROCESSING SYSTEM PAY FOR ITSELF? IN HOW MUCH TIME?)

After writing headlines, the next improvement needed for the multiperson outline is to standardize the components in the document and thus be able to map outline entries onto well-defined bits of text. This step is a bit more daring because it requires you to change not only the style of your headings but also the "look" of your document. The basic technique is to define a standard size or length for each "chunk" of the document, to decide, in effect, that the proposal or report as a whole will comprise a series of sections of similar length.

This technique is new, but hardly radical. One does not need to be a systems scientist to know that any system can be partitioned into smaller systems or aggregated with others to produce larger systems. If I told you to divide your thirty-page report into ten sections of about three pages each, you could do it without much struggle. If I told you to organize it again into fifteen sections of about two pages each, you could do that as well.

Conversely, if I then told you to take this series of fifteen sections, each with about two pages, and reorganize them into five sections with about six pages each, you could do that too.

The question is not whether you could. The question is why you should bother!

Actually, I do not want you to partition a finished report at all. I want you to develop an outline this way, thinking of your long, complex, concatenated report or proposal or manual as a series of standard bits—even "bite-sized" bits if they are small enough. Moreover, when you outline, I want you to have *one headline for each bit.* So, if you know that an average bit is, say, three pages, and you know that your outline has ten headlines, you know that the final product will have about thirty pages. Further, the person in charge of writing the text for five of those headlines not only knows what he or she is supposed to prove (the headlines show this) but also knows how much must be written.

Again, though, you may ask *why?* Why should anyone want to break up the continuous flow of words, pictures, and numbers into discrete bits and chunks? (In physics it is called "quantizing.") Further, why would any sane person insist that all the parts be of a single, standard size?

The answer is simple: Documents assembled as a series of standard-size units are vastly easier to plan (and replan), manage, write, revise, produce, read, and evaluate. Further, when each of the standard-size units is linked to an evocative, interesting headline, the result is probably the most interesting and accessible technical document possible.

The third improvement in your outline is to integrate pictures and text. Thus, your outline will include *headlines,* with a *standard-size section* to accompany each headline, and a *sketch of all the artwork that will go into that section.*

Because pictures, charts, and tables are the most important part of much technical communication (in fact, many texts about "technical writing" spend more time on graphics than on sentences), it follows that the clearest way to plan a section is to rough out the diagrams and tables for each section. So, if you know the approximate length of a section, and you know how much space you need for artwork (keeping in mind that artwork can be "shrunk" somewhat), you can deduce how many pages (or how many words) of text you need for each section or unit.

When a document is planned and assembled this way, as a series of standard-size, self-contained bits or units or chunks, it is almost always referred to nowadays as a "modular" report or proposal. This use of aerospace electronics terminology is largely attributable to the fact that this approach was first codified and popularized by a team of editors at Hughes Aircraft in the early 1960s. And the particular modular format devised by Hughes has itself been adopted throughout large segments of the aerospace and defense R&D community (and adapted with a hundred minor variations). This particular format (which I shall describe later) is now so widely used that there is a whole generation of engineers and scientists who know of no other way to prepare a proposal!

Another aspect of Hughes's pioneering work in this field is that the Hughes team of writers described it with a name borrowed from Hollywood's movie studios—"storyboard"—and gave it currency throughout technological industry.

In film making, a storyboard is a series of sketches with a bit of text beneath them (or, sometimes, beside them). In planning a film, especially an animated cartoon, the writer or director will draw each of the shots or scenes in the film, with a few words to describe what happens or what the performers are saying in the scene.

Today, sophisticated editors and "publications engineers" use an analogous technique to prepare major proposals, manuals, and reports.

The first task in storyboarding is to develop a conventional, descriptive outline. Consider, for example, this proposal to install a "zero-based" planning and budgeting system into a hospital:

- Introduction
  - Inflation of Health-Care Costs
  - Cost Containment Legislation
  - Voluntary Cost Containment Program
- Approach
  - Zero-Based Budgeting for Hospitals
  - Zero-Based Budgeting Process
- Implementation Plan
  - Forms Development
  - Staff Development and Training
  - Accounting Software

To repeat, if you were an expert on this subject and were going to write a brief, ten- to fifteen-page proposal all alone, you might need no more outline than this. But if you were about to undertake a more complex management task, you would need a more detailed, "modular" approach. You might want to prepare a *storyboard.*

The next step, then, is to "quantize" your outline into equal-size modules (let us say two or three pages each). Look, first, at your most eastward entries; start with "Inflation of Health-Care Costs." Can you write everything you want to write on that subject in two or three pages? You cannot? You want four to six pages? Then what are the two or three "modules" under that heading, and what headlines will you follow?

As a hypothetical case, suppose you break it apart this way:

Introduction
  Inflation of Health-Care Costs
  - Cost increases due only to inflation, 1945–79
  - Cost increases due to improved health care, 1945–79

Thus, you have decided to partition "Inflation of Health-Care Costs" into two modules.

Your next question is to decide whether you want any text at all under the heading of "Inflation of Health-Care Costs," whether you want to preview or introduce the discussion in the subordinate modules. If the answer is yes, you do want text, than you must decide what two or three pages of text should go with that introductory section, and what headline you should use, such as "Why Health-Care Costs Have Grown So Fast Since 1945."

Thus, each item in your first outline must be exploded into distinct modules; or the item itself must be recast as a module; or, in a few cases, two or more "shorter" items in your original outline can be aggregated into one larger module. Figure 4-2 shows how the "Zero-Based Budgeting Proposal" is "quantized" into standard-size modules.

(Incidentally, this restructuring of the outline can be done by you—if you know enough about the subjects—but is usually done by the writing team.)

Once you have your list of "modules" and a headline title for each, you prepare a small plan for each module, like that shown in Figure 4-3.

Notice that the plan has five critical elements:

1. The categorical location of the module in the outline
2. The *headline*—in capital letters
3. A *summary passage* (sometimes called the *thesis* or *thesis sentence*) in which you present a concise, compressed version of the argument, theme, assertion, or issue you will present in the module
4. A few notes to guide the text
5. A rough sketch of the artwork

FIGURE 4-2 Quantized Version of the Outline

In some companies, these module descriptions are written on 4-by-6-inch or 5-by-8-inch index cards. Some firms have large 11-by-15-inch or 11-by-17-inch preprinted forms on which to fill in the information.

Whatever the format, the next step in storyboarding is to begin to tack the module plans to a wall or large display, to put them in the order they will have in the text. (Remember, no matter how many levels in your outline, your presentation will always be one dimensional, one item after another.) In this form, the manager and the writing team review the design and "readability" of the document. They check for logic and flow; they look for materials that are out of place or for gaps that need closing. In the case of a proposal, they review all the specifications in the Scope of Work or RFP (Request for Proposal) to be sure that every requirement will be met clearly in one module.

If they do not like what they see, they have several options: remove modules, rearrange modules, add modules. Any new module is planned and sketched and is then tacked on the board with the others. In this way, a major

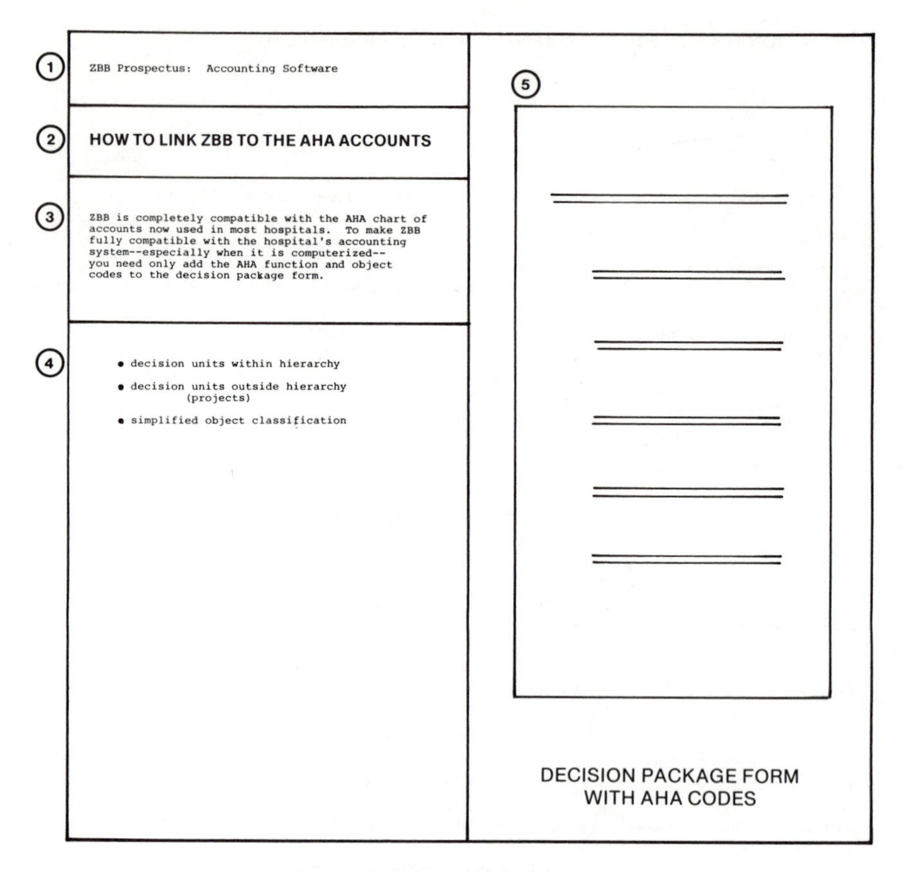

FIGURE 4-3 Plan for a Module

document can be "revised" several times in just a few hours—at just that point in the process when revising is quickest, cheapest, and most clearheaded. (Any last-minute revisions will be made, of course, by people who are bleary-eyed and exhausted.)

The particular module format pioneered by Hughes Aircraft, which has become widely used throughout American industry, has the following characteristics:

1. It is two pages long.
2. It appears on *two facing pages,* starting on the left, so that every time the reader turns a page he or she is at the beginning of a new module.
3. Typically, it will contain text on the left and graphics on the right (although there are variations on this pattern).

Figure 4-4 contains a Hughes-styled module created for our hypothetical Zero-Based Budgeting proposal.

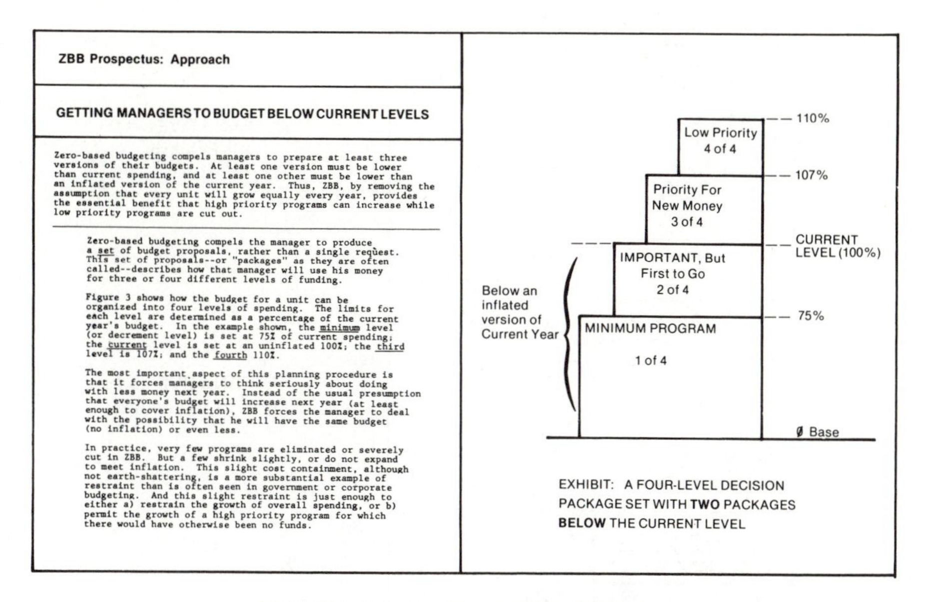
ZBB Prospectus: Approach

GETTING MANAGERS TO BUDGET BELOW CURRENT LEVELS

Zero-based budgeting compels managers to prepare at least three versions of their budgets. At least one version must be lower than current spending, and at least one other must be lower than an inflated version of the current year. Thus, ZBB, by removing the assumption that every unit will grow equally every year, provides the essential benefit that high priority programs can increase while low priority programs are cut out.

Zero-based budgeting compels the manager to produce a set of budget proposals, rather than a single request. This set of proposals--or "packages" as they are often called--describes how that manager will use his money for three or four different levels of funding.

Figure 3 shows how the budget for a unit can be organized into four levels of spending. The limits for each level are determined as a percentage of the current year's budget. In the example shown, the minimum level (or decrement level) is set at 75% of current spending; the current level is set at an uninflated 100%; the third level is 107%; and the fourth 110%.

The most important aspect of this planning procedure is that it forces managers to think seriously about doing with less money next year. Instead of the usual presumption that everyone's budget will increase next year (at least enough to cover inflation), ZBB forces the manager to deal with the possibility that he will have the same budget (no inflation) or even less.

In practice, very few programs are eliminated or severely cut in ZBB. But a few shrink slightly, or do not expand to meet inflation. This slight cost containment, although not earth-shattering, is a more substantial example of restraint than is often seen in government or corporate budgeting. And this slight restraint is just enough to either a) restrain the growth of overall spending, or b) permit the growth of a high priority program for which there would have otherwise been no funds.

FIGURE 4-4 An Illustrative Module

The basic format is text-on-the-left/graphics-on-the-right, but there are also numerous variations. Figure 4-5 shows some alternatives. Notice that every part of the module has appeared earlier in the original storyboard plan except for the explanatory text on the left-hand page. There are, therefore, almost no surprises in the finished product.

Indeed, this lack of surprises, this extraordinary control of the finished product, is the principal advantage of the modular approach to outlining. It makes the job manageable, levels the effort, and reduces the overall impact of intermediate delays. It does for the writing process what morphological analysis does for the industrial engineer: gets complicated things under control. It even permits sections of the document to be page-numbered and printed before the entire document is complete—a considerable timesaver on a tight proposal production schedule.

And the evidence is that readers also approve heartily. (I would not recommend the approach if it would not increase your effectiveness.) Readers of technical materials, particularly readers of competitive proposals, genuinely like to have long and complicated material reduced to short, stand-alone units; they like to read headlines that tell them where the specs are addressed, spec by spec. They like to be able to look at a table and the accompanying text *without needing to turn a page.*

But are there any shortcomings in this approach? Is it a panacea for the planning and organizing of all complicated technical materials?

In one form or another, this modular approach will help the planning and outlining of almost every long document you are likely to write. Even if the

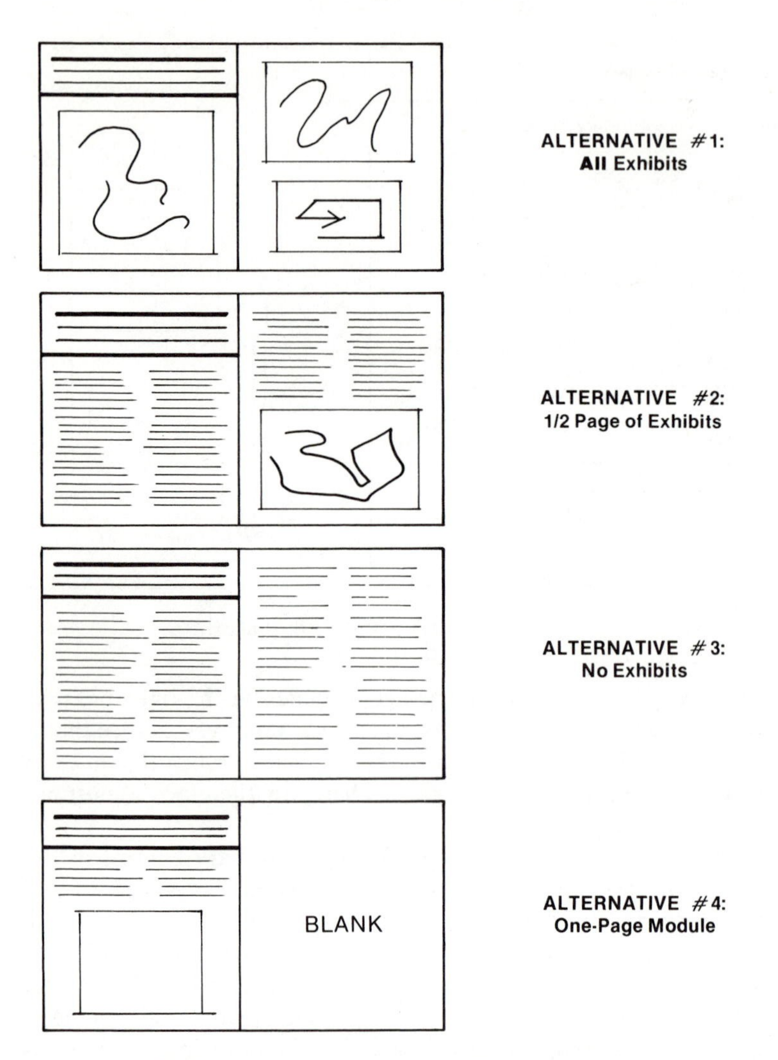

FIGURE 4-5 Variations on the Basic Format

orthodox Hughes-type module is inappropriate, the features of the approach can be applied where suitable:

- Almost every professional communication benefits from having headlines instead of headings.
- Almost every difficult document benefits from having short sections with lots of headlines.
- Almost every technical message benefits from having a summary or thesis passage at the beginning of each section.

- Almost every document that uses graphics and tables benefits from having the accompanying text immediately adjacent to the figure or chart.

By adapting the approach—changing the length or format of the modules—you can handle the complaints of those writers who find the discipline too Procrustean. In fact, the only serious obstacles I have run across in the use of the Hughes type of facing-pages module are, first, the inability of some writers to use pictures and exhibits well (so that their modules contain a lot of gratuitous cartoons), and, second, a reluctance to print or copy on both sides of the paper.

(The technique also tends to increase, slightly, the number of pages and quantity of artwork.)

These objections notwithstanding, the "modular" or "storyboard" approach is unquestionably the best way to produce a detailed outline for a team of writers (Step 1.12).

It is also the best way to secure your supervisor's approval of the plan (Step 1.13), if that is called for. To be sure, a storyboard plan will give the clearest preview possible of the finished product to your supervisor (or to the "red team" if they can be convinced to review the document in the planning stage). Further, if the supervisor does not approve of the plan, he or she will more easily be able to identify precisely what needs to be changed or improved, so that the impasse can be resolved quickly.

At the end of Chapter 3 I urged you not to be intimidated by the strategic-planning process. Perhaps it is time for another assurance.

The outlining techniques described in this chapter can be done in as little as five minutes or as much as five days, depending on the scale of the message you are planning. If done well, outlining will almost always *save* you time in the end. (In the case of rushed proposals, good outlines can even save your health!)

When my clients ask me how long this should take, I tell them that the entire process of strategic planning and outlining, the whole planning phase should take about 10 percent of the time they would ordinarily have spent on the whole project (including revisions). Thus, if they would ordinarily have spent one hour, planning should take about six minutes; one day, about forty-five minutes; one week, about half a day.

Notice, though, that I describe the estimate in terms of the time they would *ordinarily* have spent. This "ordinary" time should shrink, so that there is a net saving in the total effort.

To make your planning go more efficiently, you should know that there are certain typical design problems in the most common types of business and technical writing. The next chapters give some practical advice on the best ways to design reports, proposals, letters, and other projects that often come up in business and technical communications.

DESIGN GUIDELINES

5

Designing Reports

CHARACTERISTICS OF "GOOD REPORTS"

A *report* is a message written to fulfill an assignment. In some cases, writing the report *is* the assignment—or the last task in the assignment. In other cases, the report "writes up" the assignment, proving that it was, in fact, fulfilled.

Reports, therefore, are as diverse as the assignments performed by engineers and scientists: progress reports, status reports, trip reports, inspection reports, problem-solution reports, reports for publication. Reports can be in any format: long or short, letter or memo, formal or informal.

Can any single set of criteria be applied to such a motley collection of documents? Are there particular characteristics associated with every good report?

Surprisingly, nearly all good reports have certain common traits. And a fairly good report that lacks one or more of these traits can always be improved by making the appropriate change. Thus, a good report will

- Fulfill its assignment
- Begin in a way that engages the reader

- Be organized according to a logic that the reader can perceive
- Move forward in a continuous flow
- Separate facts from opinions
- Be free from excessive detail
- Be "just right" in length
- Use the least-formal "style" acceptable
- Observe good communication manners
- Make intelligent use of exhibits
- Provide frequent guides and aids for the reader

A good report fulfills its assignment. Whether the task was assigned by the author's boss or sponsor, or whether the author had his or her own mission in mind, a well-designed report *must* fulfill that assignment.

Typically, this criterion is applied by the author's superiors; often the "acceptability" of a report is a provision in a research and development contract. So, a writer who wants to be effective—and contain the time it takes to prepare reports—will learn exactly what criteria will be used to judge the report acceptable. For example, will the reader demand an itinerary as part of the trip report? Will the sponsor want the "raw data" that support your conclusions? Will the vice-president want your recommendation, or just your comments on the alternatives?

A lab manager in a consulting firm once told me that the worst thing a consultant could do was submit a final report that was *useless* to the customer. He was right; whatever other virtue your report may have will be obscured by its lack of usefulness or acceptability to the sponsor. There will be no consolation prize for neatness and spelling.

A good report begins in a way that engages the reader. In fact, a report that begins badly rarely gets better. (And even if it does, few are likely to read it.)

There is no more effective way to begin a report than with a statement of your own objective. Remember, your objective expresses your intention to have the reader know something (informing), believe something (persuading), or do something (motivating) as a result of your communication. So, there is no more engaging opening (think of the way gears engage) than to state exactly what you want from the reader, or what you intend to do to the reader.

Do *not* begin with history or background. Begin with your objective. Do not begin with

> Until July of this year, the design of the X104 engine has proceeded ahead of schedule.

Instead, write

> The purpose of this memo is to warn you that we will probably be two months late in completing the prototype model of the X104 engine.

The second opening grabs and holds the attention of the reader because it explains why the memo was written. When the message is persuasive or motivational, the impact is even greater:

> The purpose of this memo is to convince you that the Project Management has underestimated the time for the X104 prototype by two months. (persuasion)

> You will have to meet with ABCO, Inc., and tell them that we are going to be two months late on the X104 prototype. (motivation)

When you begin with an opening like this, you are writing to an alert, awake reader. As the citizens' band radio operators put it: Your reader "has his ears on."

Put in negative terms, a reader should *never* be able to read the opening of your report and say So what? or Who cares? If there is a Summary at the beginning (curiously, summaries until recently appeared at the ends of things), the Summary should engage the reader. If there is an Introduction, the Introduction should engage the reader. The troublesome cases are those short reports—often in the form of memos—that have no sections or headlines. In these the first paragraph must function as a summary, as a preview of the material most important to the reader.

A good report will be organized according to a logic that the reader can perceive. A well-written report is never a mystery. The readers know—from the very first sentence—why they have been asked to examine this report, that is, what exactly they are expected to know, believe, or do after reading it. Moreover, the order of presentation must make sense; the readers must be able to perceive it. Are the materials organized by time or sequence, by logical function, by the parts in a proof, as a scientific investigation? If the order or logic is not self-evident, then the report must contain a preface or foreword that explains how it is organized.

The most serious organizational flaw in the writing of scientists and engineers is a tendency to follow the chronology of events: background-problem-method–of–solution-data-interpretation-conclusions-recommendations. In almost every case, the most engaging part of a report is its Conclusions and Recommendations. Those firms that require the Conclusions and Recommendations to appear *first* know what they are doing; they are starting with the material that will most interest and involve the reader.

A good report moves forward in a continuous flow. This characteristic has two aspects: *forward* and *continuous.* No good report sends its primary readers skittering forward and backward with references to earlier and later sections. (Of course, it is all right to assure a reader that you will return to a subject later—provided later is a better time for it.) Most people will not follow a report that is organized like a scavenger hunt, with references and citations that point like clues. (The only documents that sometimes work well this way are "programmed texts," and there are many readers who cannot handle even these.) A good report starts on the first page and keeps moving forward.

Also, a good report moves *continuously,* without interruption. Readers will not leap forward several pages and then struggle back to their original place. In fact, most readers will not finish a sentence that requires them to turn more than one page (because of interrupting tables or exhibits). Most readers will not even flip to the end of a section to read a "footnote," or jump ahead to examine a table or figure.

Do not misunderstand. This is not to say that the typical reader starts at the beginning of a report and keeps reading continuously until the end. Hardly. Most readers skim and skip, move forward and backward on impulse. *But no reader can be expected to follow a report that requires him or her to skip and lurch.*

(Notice how many of these unacceptable techniques were mentioned earlier in the discussion of writing for the "secondary audience." The point is, precisely, that only a secondary reader should be inconvenienced this way; for the primary reader the report should be a "straight shot," without detours and misdirections.)

Recall that in the Hughes Aircraft *modular* format, the text and pictures are always visible simultaneously, usually with text on the left and pictures on the right. The reader is almost *never* referred to a graphic or table that cannot be seen at the moment of the reference. How different from some firms' practice of placing all the tables and charts at the end of the report—thereby guaranteeing that most of them will never be studied intelligently! (In my own work I would sooner repeat a chart or table than send the reader scurrying off to another section in which the exhibit I am discussing was first used.)

Good reports will always separate facts from opinions. Successful writers will never allow hard data and soft judgments to appear in the same paragraph or list unless clearly identified as such. Indeed, the worst blunder that report writers can make is to try to pass off an opinion or conclusion as a fact. Not only will they appear devious and dishonest but even their hard facts will become suspect.

There should be no mixing of materials—accidentally or purposefully. It is also useful to separate your various opinions according to your degree of confidence in them. Keep the reasonable judgments away from the not unreasonable speculations; keep the educated speculations away from the wild guesses. Do not allow your worst material to contaminate your best—especially when there are critics looking for faults in your work.

A good report is free from excessive detail. Two planning errors cause excessive detail. First, failure to define clearly your objective causes you to provide material *irrelevant* or *unimportant* to your aim. Second, failure to analyze your audience causes you to add background and explanation that may be unneeded.

If you appraise your "raw material," as I described earlier, you will be better able to decide what information should be featured, what should be left out, and what should be placed in attachments at the end—where almost no one will read it. And even when you decide that a particular table or a complicated exhibit should appear in the body of the report, be sure that only the relevant portions of the table are in the body by devising a "secondary table," like that shown in Table 5-1.

A good report will be "just right" in length. Knowing the appropriate

TABLE 5-1 Secondary Table Derived from Long Table

Primary table, showing more information than is relevant.

Recent Trends in Expenditures for Science and Technology (1960–1976)

	1960	1965	1970	1974	1975	1976
Federal space program outlays (billions)	.9	6.9	5.5	4.9	4.9	5.3
NASA outlays (billions)	.3	5.0	3.6	3.0	3.0	3.3
R & D Funds (billions)	13.5	20.0	25.9	32.3	34.6	37.4
Basic research (percent)	8.9	12.7	13.8	12.7	12.9	12.8
Federal funds (percent)	64.6	64.9	56.6	52.3	53.0	52.9
Industry funds (percent)	32.7	32.2	39.7	43.6	42.8	42.8
Defense-related outlays (percent)	52	33	33	29	27	26
Space-related outlays (percent)	3	21	10	7	8	8
Federal obligations for R&D (billions)	7.6	14.6	15.3	17.4	19.0	20.7

Secondary table, showing only that information under discussion.

Relationship between NASA Expenditures and the Federal R&D Budget (1960–1975)

	1960	1965	1970	1975
Federal obligations for R&D (billions)	7.6	14.6	15.3	19.0
NASA Outlays (billions)	.3	5.0	3.6	3.0
Ratio of NASA outlays to all Federal R&D (percent)	3.9	34.2	23.5	15.8

length of a report is difficult. In general, reports should be as short as you can make them—but there are many exceptions.

Excess length creates different, hard-to-predict effects. Most people believe that great length is a sign of completeness and thoroughness—the longer the report, the greater the evidence to support the conclusions. Others, however, construe length as *defensiveness* or *laziness* or even *arrogance.* (One easy way for an uncooperative technical person to answer a question from a "layperson" is to throw together a large intimidating pile of information.)

On the other hand, shortness can be taken as a sign of confidence and conclusiveness, but, unfortunately, also as evidence of *carelessness, incompleteness, superficiality,* or *curtness.*

Notice that these effects of length have little to do with the actual content. Rather, each report should be planned to have a certain "heft." Slight and superficial design changes—like the layout of pages, the style of type, or even the thickness of the paper stock—can produce the effect you want.

Or, when more substantive changes are needed, you may have to trim away or restore materials that your planning judged marginally important. That is, you can either relax or tighten your standards on what is relevant to your audience and purpose. If you need more, err on the side of of "too much." If you need less, err on the side of "too little."

A good report uses the least-formal "style" acceptable for the occasion. Good report writers do not show off; they do not "overdress" for the occasion.

To a large extent, most errors of "style"—using wordy, windy, difficult expressions—are corrected later in the editing stage. They are not, strictly speaking, a problem that can be controlled in the design stage of the report. Except that I know engineers and scientists will write a better report if they refrain from putting on a stiff pompous personality—even during the planning stage.

This is *not* to say that you should write the way you speak: Speaking is a radically different approach to language, and most of the rhythms and forms of speech would not work in your formal reports. But you *should* try to write more like a person, and less like a robot or an institution.

Further, **good reports observe good manners.** Well-written reports hardly ever contain sarcasm or insult; they are free from harsh, abusive language and gratuitous criticism. They do not flagrantly offend or assault the reader's beliefs.

I have yet to read a good report written while the writer was steaming with anger or trembling with fear. (I routinely throw my angry reports away—feeling better for the exercise.)

Unfortunately, these last two criteria sometimes conflict, confusing the writer. Is it unmannerly to use slang or contractions in a report? Is it unseemly to express any emotion in a business report other than "strong concerns"? Does a formal, polite communication almost force you to be stiffer and fancier than you would like to be?

In general, when you know your audience well you are safer to err on the side of *less* formality. When you do not know your audience, however, you are better to err on the side of politeness.

But never, never write in that bloodless, unreadable style so often associated with scientific papers or government regulations. Never write

> Due to uncertainties in the anticipated duration of the truck transport stoppage situation, it has been found impossible to ascertain when your equipment delivery schedule will become operable.

when you can write

> Until the trucking strike ends, we will not be able to promise a delivery date for your equipment.

Good reports make intelligent use of exhibits. Exhibits—pictures, diagrams, charts, tables—give vitality to a report. Almost every problem, process, or array of data is clearer in an exhibit or graphic presentation. It is not uncommon, indeed, to plan a major report as a series of exhibits with accompanying text.

Finally, **a good report provides frequent guides and aids for the reader.** In other words, a good report has an external skeleton: Everyone can see how it has been built. (Literary essays, on the other hand, tend to have internal skeletons; you must actually study them to divine how they were composed.)

A good report, therefore, is filled with devices that guide the reader through the text:

- Headings and headlines—the more the better
- Numbers, letters, and "bullets"—to put ideas into lists
- Previews, summaries, and introductions at the beginning of the report and at the beginning of each major section—to alert the reader to what is coming
- Reviews and recaps—to remind the reader of what was just established
- Typography and graphic embellishment—to highlight and set off key passages; marginal glosses—to explain what is covered in each section or paragraph
- Wide margins, large print, and shorter paragraphs—to make the work more accessible

In short, anything you can do to guide the reader through the elaborations of your thought. Notice the "before and after" specimen in Table 5-2.

These then are the criteria that separate well-designed and well-presented reports from the average. They apply to reports as a whole, or to any component of a report.

TABLE 5-2 The Effects of Graphic Embellishment

Before: No Graphic Embellishments

To: T.K. Jones

Re: Safety of Copier Chemicals

In its reply to your inquiry about the potential health hazards of copier chemicals, the Division of Employee Health and Occupational Safety says that there is no evidence linking these chemicals to any illness. The Division does recommend, however, that copiers not be placed very close to employees and that all copier areas be well ventilated.

The Division also makes the following points. First, the toxicology literature shows that no disease is associated with exposure to dust from dry-ink toner. Although most dry-ink toners contain small amounts of carbon black—a chemical containing carcinogenic benzene extracts—these extracts are harmless within the carbon black compound.

A recent report suggests that dry-ink toner can cause mutations in certain bacteria. Until further testing, though, it is impossible to know whether this finding affects humans. Further, both the Xerox and Nashua companies have begun to eliminate the most suspect chemical, nitropyrene, from their inks.

The EPA states that "based on information currently available, the public should not be alarmed about the safety of using their present copiers or current copying practices."

After: A Few Graphic Embellishments

To: T.K. Jones

Re: SAFETY OF COPIER CHEMICALS

In its reply to your inquiry about the potential health hazards of copier chemicals, the Division of Employee Health and Occupational Safety says:

1. There is *no evidence* linking these chemicals to *any* illness.
2. Even so, copiers should *not be placed* very close to employees.
3. All copier areas should be *well ventilated.*

The division also makes the following points:

First, the toxicology literature shows that *no disease* is associated with exposure to dust from dry-ink toner. Although most dry-ink toners contain small amounts of carbon black—a chemical containing carcinogenic benzene extracts—these extracts are harmless within the carbon black compound.

TABLE 5-2 *(cont'd)* The Effects of Graphic Embellishment

After: A Few Graphic Embellishments

Second, a recent report suggests that dry-ink toner *can* cause mutations in certain bacteria. Until further testing, though, it is impossible to know whether this finding affects humans. Further, Xerox and Nashua companies have begun to *eliminate the most suspect chemical, nitropyrene,* from their inks.

The EPA states that, based on information currently available:

> "THE PUBLIC SHOULD NOT BE ALARMED ABOUT THE SAFETY OF USING THEIR PRESENT COPIERS OR CURRENT COPYING PRACTICES."

THE COMPONENTS OF A REPORT

Most complete reports have five parts:

1. Front material
2. Beginning
3. Middle
4. End
5. Attachments

The *front material* contains the title (or title page), the identification of the author, tables of contents and exhibits, acknowledgments, foreword, preface, abstract, summary—everything necessary to capture the interest of the readers and convince them to read the longer document.

At the other end of the document is a set of *attachments:* appendixes, supplements, glossaries, reference materials—anything that expands or "backs up" the message but is not integral or essential to the presentation.

Between the front material and the attachments is a section often referred to as "the body." I prefer to call it "the story" because, as it turns out, good reports always tell a story. Informative reports describe interesting facts and processes; persuasive reports lead to a provocative or intriguing conclusion; motivational reports spur you to action.

The story section of a report has three main parts, known technically as the beginning, the middle, and the end. (Yes, those are the technical names.) What I have discovered, however, is that despite the diversity of reports, and despite the seemingly thousands of forms they may take, the beginning, middle, and end of a report are usually about the same things:

- **The Beginning** explains why the report was written, or why the work that led to the report was undertaken.
- **The Middle** tells how the work was done.
- **The End** offers the results or reward of the work.

Even though all reports have this basic structure in common, there are also differences associated with the three functions or purposes of reports. As Figure 5-1 shows, the three basic parts of an *informative report* are

- **The Beginning** — Some question of fact that the writer wanted to answer; it can be a scientific fact or a fact about the progress of a project, or the answer to any other question that can satisfactorily be answered with numbers, yes-or-no, or straightforward assertions about what is true.
- **The Middle** — A method or procedure suitable for collecting the data or other information needed to answer the initial question.
- **The End** — An answer (or answers) to the question, offered without much embellishment or justification: just the facts.

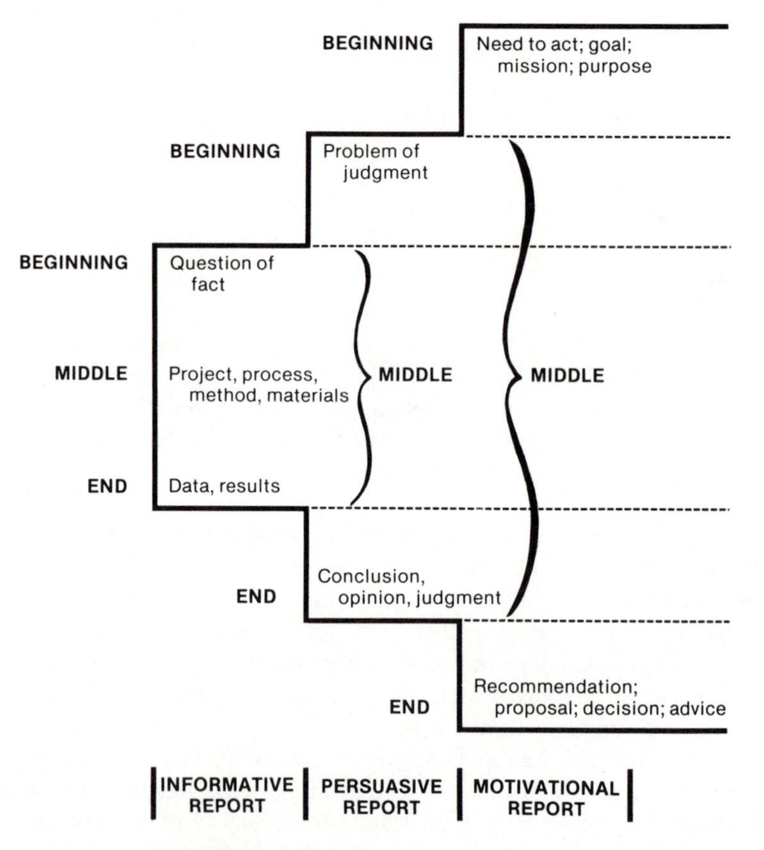

FIGURE 5-1 The Structure of Reports

The *persuasive report* has the same general parts, but the functions change. In the persuasive report the components are

- **The Beginning** — Some more-complicated question of judgment, a "problem" in the scientific sense, a question so complicated that one needs expertise not only to answer it but even to know how to go about finding the answer; the question cannot be answered with facts alone.

- **The End** — Judgments, conclusions, inferences, opinions, which address the question raised initially.

Notice that I included no discussion of The Middle of the persuasive report. That is because *the middle of a fully-developed persuasive report is the informative report.* In other words, a report that begins with a complicated problem (persuasive) will typically proceed to several questions of fact necessary to approach the more difficult problem.

For example, suppose the problem were: What is the most cost-effective way to manufacture gasohol? There would then be a series of factual questions related to this broader problem, questions about the available processes, the costs for acquiring and operating those processes, and so forth.

Or the problem might be: What are the health risks associated with taking annual chest X-rays? This problem would generate dozens of factual questions about the kinds of equipment in use, the amount of radiation they emit, the scientific consensus on the health effects of those emissions, and so forth.

Once the more-complicated problem has been "decomposed" into factual questions (sometimes referred to as the hypothetico-deductive method), the report proceeds to describe processes, experiments, methods, and other items that ordinarily appear in the middle of the informative report. Then the actual facts and data are reported, as in the informative report.

Thus, the persuasive report, in its most sophisticated form, ends with conclusions drawn from the data. (It must *not* end with the data alone, or it is not a persuasive report!) The informative report is, in effect, nested inside the persuasive report, so that the entire series of components in the persuasive report will often be:

	• Problem ↓
INFORMATIVE REPORT	• Questions of Fact (research questions, hypotheses, etc.) ↓ • Method ↓ • Results ↓
	• Conclusions

Also shown in Figure 5-1, the *motivational* report has still a different beginning and end:

- **The Beginning** — A need or impulse to act, a sense of mission, a goal, a decision to decide, an intention to purchase, and so forth; unlike the persuasive report, the motivational report begins with a real stress and urgency about what to *do,* not just what to think or expect.

- **The End** — Recommendations, proposals, and exhortations to act, buy, move, select, decide.

The middle of the motivational report is, as you no doubt have guessed, the entire persuasive report.

In the most familiar logical model, the need to act (motivation) leads to the formation of complicated research problems (persuasion), and then the whole process begins again. For example:

We must reduce our use of oil.	________	(Need to act)
How soon can we set a minimum standard of 30 mpg for passenger automobiles?	________	(Problem of judgment)
What are the fuel consumption rates for automobile models currently in production or planned?	________	(Question of fact)
or		
We must reduce mortality from lung cancer.	________	(Need to act)
How can we reverse the increase in smoking rates for young adult women?	________	(Problem of judgment)
What are the results of our study of techniques for stopping smoking?	________	(Question of fact)

So, the entire series of components in the motivational report will often be:

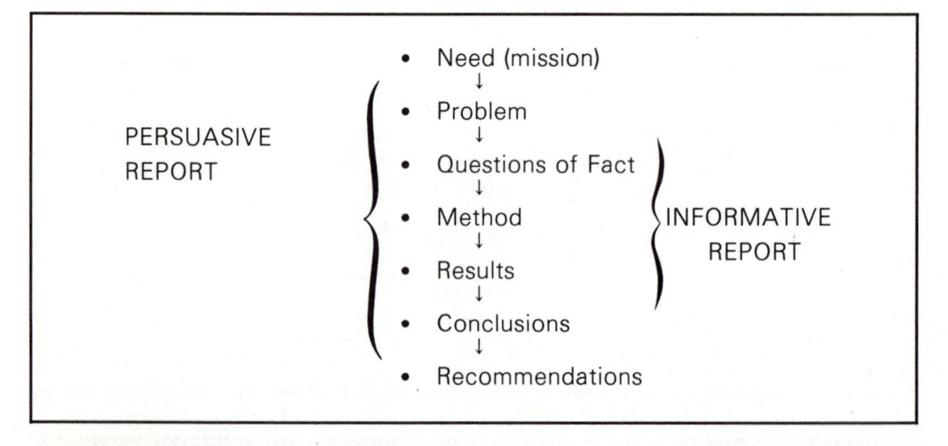

Sometimes the distinctions among the three kinds of reports are difficult, but you will usually see that your report fits best into one of the categories. In short, the kind of report you are planning is a function of how it begins and ends: questions of fact—informative; problems of judgment—persuasive; need to act—motivational.

Do not misunderstand. Not all motivational reports have the structure above (nor do all persuasive and informative reports look like those I have discussed). There are motivational reports that contain no data; there are persuasive reports that do not describe their methods; and there are informative reports that give answers without the questions.

Nevertheless, the structure described in Figure 5-1 is the logical foundation for all scientific and technical reporting, and for the best management and governmental reporting as well. If you decide to depart from this structure as you design your report, you should do so consciously, for a reason. Failing this, you may write a report that only you can follow, or a report that fails to communicate its own objectives to the reader, or—what is always a serious possibility—a report that will be demolished by your critics and competitors.

Ultimately, a well-designed report is one that fulfills its assignment (either the boss's or the writer's) in the face of the natural resistance that greets all reports. Every report has inherent enemies, forces in the world that work against its success. For the informative report, the enemy is *entropy,* the natural force of noise and disorder. For the persuasive report, the natural enemies are *prejudice* and *fallacy,* easy and careless ways of forming conclusions and testing beliefs. For the motivational report, the natural enemy is *inertia,* the natural tendency in your reader to do nothing or to resist change.

If you want to engineer an effective report, then, you will have to design it so that it can compensate for or overcome its natural enemies.

How to Make a Report Inform

Information is the reduction of uncertainty. When we are uncertain or confused about what is true, information is the stuff that removes our uncertainty and dispels our confusion. When we cannot predict future events we care about, the knowledge of those events, as they occur, relieves our anxiety.

The people who study information formally are usually communication engineers, and what they study is the statistics of uncertainty: the relationship between probability and information, the differences between information and noise, the construct of "channel capacity," and related issues.

But a few of these scholars have also been able to apply these same mathematical models—used mainly in the design of electronic communication systems—to ordinary *human* communication through speech and writing. As a result, we can offer some added scientific rationale for those techniques that good editors and artists have always used instinctively.

To be informative, you must *separate your information from the chaos, noise, and other information in your reader's environment.* You should presume that there is a natural tendency for your reader *not* to receive your message, *not* to read it once received, *not* to pay attention to it once read, *not* to interpret it accurately once attended to, *not* to store or remember it once interpreted. This natural tendency for your message to become lost in the chaos of your reader's environment, to degenerate from meaningful information into meaningless letters and numbers, is often called *entropy* (borrowing a term from thermodynamics).

On one level, entropy may be thought of as a purely physical phenomenon: Pages can get lost, destroyed, shuffled out of sequence; sentences can get garbled going through the voice or facsimile telephone channels; message carriers go to the wrong destinations or mix your message with others. (Murphy's law—If anything can go wrong, it will—is a description of entropy.) In short, the most likely career for any physical object, including messages, is disintegration; only deliberate intervention can affect that.

At another level, entropy is also a psychological and physiological phenomenon. Except in special cases, readers do *not* necessarily want to read, they do *not* want to exert themselves, they do *not* want to learn, they do *not* want to understand. Indeed, there are times when students of communication must wonder that anyone ever gets any message right!

This problem, the need to battle and overcome entropy and distraction, is fundamental in all communication. In an informative report, it is almost the entire issue. Thus, for an informative report to be effective, it must

- Capture and recapture the reader's attention
- Guide the reader's eye
- Assert understandable claims and facts
- Suppress excessive and irrelevant details
- Build an apparent, predictable structure
- Balance general and abstract statements with illustrations, samples, and examples
- Use pictures and exhibits, in appropriate places

To *capture and recapture the reader's attention,* you must announce your intentions from time to time. Of course, your opening should state your purpose —or use some other material that is quite likely to engage your reader. Moreover, each major section should announce its sectional purpose as well. Use previews and forewarnings; and when the sections are long or the material complicated, use wrap-ups and summaries (in the older sense of the word, a recapitulation of the latest section).

To *guide the reader's eye,* be aware of visual boredom and fatigue. Distrust "sameness" in your report or document. Think up ways to use graphics, italics, boldface, underscoring, occasional question marks. Indent some sections. Use lists with "bullets." Refrain from asking your reader to look at two things at once (text here, diagram there).

Make your claims understandable by testing them with an intelligent, naive reviewer. That is, a reader who knows as much as your intended audience but does *not* already know what you mean. Do not overqualify or hedge your claims.

Also, do not try to state every contingency and exception before you make the general claim. Watch out for statements like the following:

> We have eliminated half the incidence of the disease, but because this is a chronic disease and the annual incidence constitutes only 5 percent of the total prevalence, and because our serum does not affect existing cases of the disease—except for about 25 percent of those in which the onset is before twelve years of age (provided the child has no other history of respiratory disease)—we predict, using two-year-old data as the basis for our forecast, a drop of only 2.1 percent in the prevalence indicator.

Suppress excessive and irrelevant details by thinking of your objective and audience. Does the material you are about to write help you to achieve the precise informational objective you set for yourself? Will it help you to make your message more interesting or accessible to the reader? If not, why are you about to write it? At the very least, picture your reader to see if he or she wants or needs the paragraph or table you are about to compose.

To *build an apparent, predictable structure* you must, first, have some logic or purposeful sequence in your plan. Then, you must explain it to your reader —several times if it is complicated or novel. You must tell the reader explicitly that the results are presented *in order of importance,* or *in the order in which they were collected,* or even *in alphabetical order,* if that is all there is to it. You must explain why one section follows another. Write: "Before evaluating the structural soundness of this fabric, we should first describe the unusual series of tests to which it was subjected." Or: "Part 3 describes each of the components in the system in more detail." Or: "Those curious about the details of the simulation model can find them in Appendix 2."

Always tell the reader why one thing follows another—especially when the document is long, but even when it is short. In a one-page memo reporting five problems in a new design, you should tell the reader what the order of presentation of the five items signifies. Even if the order is random or accidental, you should still say so.

When an investigator or a researcher or a problem solver has worked all the way through a problem to the conclusions, he or she tends to forget that the reader did not come along on that labyrinthine journey. The document becomes too abstract and general. You must *balance abstract and general statements with illustrations, samples, and examples.* Especially when you are trying to *teach* something in your informative report, you must use cases, specimens, anecdotes, and particulars. The phrases *for instance* and *for example* are marvelously useful ways to start a sentence or paragraph. Likewise, *to illustrate* and *(let us) suppose.* If you are unsure of how many examples and illustrations you need, show your material to that intelligent, naive reader and ask this reader to make a mark in the margin everywhere he or she would like a "f'rinstance."

Finally, *use pictures and exhibits, in appropriate places,* whenever they help the reader. Be aware that readers like pictures and tables (if they are clear and well made), but typists, publications managers, journal editors, and book publishers will try to talk you out of them. Graphics cost more money, usually, than text, and most typists can do five pages of "straight typing" in the time they need for a one-page table. So, use your judgment and be sure that any point that can be made better with a chart or a table or even a photograph is made that way. If anything, go for more than you really need, knowing that all the people in the typing and production chain will manage to talk you out of a few of them.

How to Make a Report Persuade

To be persuasive you must *prove* the truth of your conclusions. I am not talking about mathematical proof (what the logician calls "demonstration"), but about the art of building a convincing case. If you set out to shape attitudes, opinions, or beliefs, you must win the assent of the reader. And, when you set out to change strongly held beliefs, you must be prepared not only to persuade but to grapple with the reader's previous notions and prejudices.

A well-designed persuasive report usually begins with a statement of the proposition to be "proved." For example:

> The purpose of this report is to show that Acme Oil Company has 40 percent more heating oil in its inventory than it reported last month.

or

> This report will demonstrate that Dr. Jones has no conflict of interest in the current contract negotiations.

or

> This study will establish that BB101 meets the agency's safety standards for antiseptic soaps.

or

> In this report I shall prove that the control panel for the X555 aircraft is unacceptably susceptible to humidity damage.

or

> This report presents the estimated costs and completion date for the Smedley Project.

Notice that all of these statements contain "problems of judgment." Not one of these problems can be treated as a matter of simple information, because not one can be discussed with facts alone. Even the first instance—disputing the factual assessment of the oil reserves of a particular company—is made persuasive because there is *controversy* and *dispute.*

The goal, then, of the persuasive report is to prove that your proposition is true and, by implication, that no other competing proposition is true (or as plausible as your own).

Now, if you ever took a course in logic or rhetoric, you may remember a little about the art of proving claims. In particular, you may remember the formal or material proofs—the syllogisms. In this style of reasoning:

1. All men are mortal.
2. Socrates is a man.
3. ∴Socrates is mortal.

Using a major and a minor premise (or sometimes more), you deduce an inescapable, logical conclusion.

This sort of formal reasoning is very rare in business or professional writing, although the style of the syllogism is sometimes used to lend an air of logical authority to ordinary arguments. In most real cases, though, writers prove their conclusions by referring to some data (their own or others') and insisting that their conclusions follow from those data.

The most important part of your proof is the logical link—often unexamined—between the evidence and the conclusion. This extra premise or idea (often called the "warrant") tells us *why* this evidence leads to that conclusion. (See Figure 5-2.)

In technical and professional writing, most persuasive messages are supported with one of five "warrants":

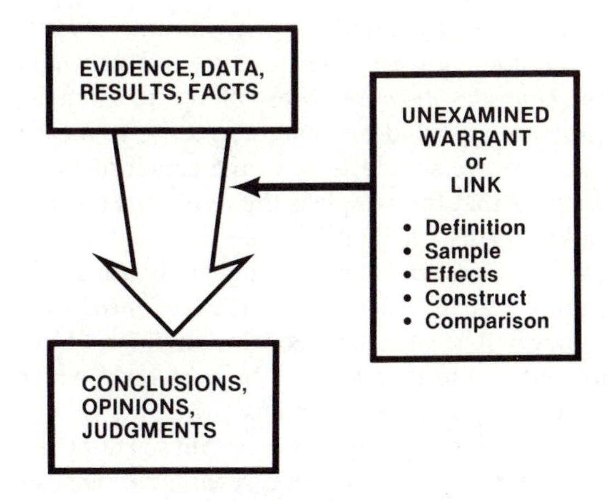

FIGURE 5-2 Use of Warrants in Persuasion and Proof

- Definition
- Sample
- Effects
- Construct
- Comparison

In proving by **definition**, there is some unexpressed definition, standard, or value judgment that underlies the proof. For example, an economist argues that because there is a shrinking money supply (evidence), the nation is entering a recession (conclusion).

Or you might prove that because your project has a .5z score (evidence), there is every reason to believe that it will finish on time (conclusion). In the first case, the economist's proof is based on a *definition* of recession (a time when the money supply is shrinking); in the second, your proof is based on a *standard* (for most projects, if $z \geq .5$, the project is likely to finish on time).

People even "prove" value judgments with unexamined standards or definitions.

For example, someone might assert that because almost no one wants to pay extra for automobile airbags (evidence), they are an unsound safety system (conclusion). The unexamined definition here is that any safety system that is unpopular or expensive is unsound.

Most scientific proof is built not on definitions but on **samples.** In this approach (often called induction or generalization), the writer insists that what is true for the sample will be true for the universe, that what is true for one or more cases will be true for all other unexamined cases. (Yes, you can prove a conclusion with a sample of one. You can prove that a general claim is false by offering one contrary example.)

The warrant for proof by samples is the warrant for all science and inference: *This sample is representative of the universe, and the rules that hold for this sample hold throughout the universe.* Most of the techniques are intended to ensure that this warrant is sound or, more correctly, to measure the degree of confidence and certainty we should have in the conclusions.

And, to the extent that the sample is the result of a representative sampling design, we generally do trust the conclusion.

We have somewhat less confidence, though, in nonscientific proofs based on samples. For example, "ABCO deliveries were late three times this year (evidence); therefore, they will be late again (conclusion)." Or, "I have had no trouble with the transmissions in the four Fords I have owned (evidence); therefore, I expect no trouble with my new one (conclusion)."

Although this last pair of proofs has a certain scientific air, you can clearly see that the warrant (that the sample is representative) is suspect and that the conclusions are hardly solid.

Still another way to prove claims is to look for the **effects of causes.** For example, "Jones changed his vote after a meeting with Smith (evidence); therefore, Smith influenced Jones (conclusion)." Or, "OMEGA, Inc., replaced its two large computers with a network of minicomputers (evidence); therefore, OMEGA must have been unhappy with its large computers (conclusion)."

The warrant behind these "arguments from effect" is that *the single explanation for an event is obvious in the event itself,* that the cause can be known from the effect, that the criminal can be known from the clues, and so forth.

The fourth common technique of proof is to reason from a **construct,** usually correlation or causation. In this case, you might argue that "because there has always been high employment during wartime (evidence of a correlation), there will be high employment if we go to war again (conclusion)." Or, "Whenever we have awarded the contract to the lowest estimate we have had an overrun (evidence of correlation), so we cannot expect to bring in a job at the cost projected in the lowest estimate (conclusion)."

Correlation and causation are so similar that people confuse them. Logicians recognized this thousands of years ago as one of the most dangerous fallacies: *Post hoc ergo propter hoc;* Because B follows A, A causes B. Despite the obvious flaw in such reasoning, even today we can read the "proofs" that use evidence of correlation as proof of causation. Many educational researchers, for example, believe that because there is such a high correlation between academic achievement and self-esteem (evidence of correlation), we can improve the academic performance of young people by raising their self-esteem (causal conclusion)!

As in the proof from samples, the proof from construct is built on the warrant that whatever holds true for the sample in the evidence will hold true for other cases in the universe.

In proof by **comparison,** the writer believes that he or she has constructed a model or representation of the universe, and that by studying the model he or she can predict or "prove" what the universe will do.

In modern engineering and science, our comparison proofs come from mathematical models, computer simulations and games, physical models, formal experiments, and many new and fascinating decision technologies.

Actually, though, our modern comparisons are derived from ancient proofs by *analogy.* "Because you wouldn't leave a fox to guard a henhouse (analogy/evidence), you shouldn't let the health-care industry police its own operations and costs (conclusion)."

Obviously, the warrant behind each proof by comparison—whether it is built on an informal analogy or a formal model—is that the comparison is valid and apt, that the model or representation or figure does, in fact, describe and predict the behavior of the universe. Because no model is identical with the object modeled (or else it would not be a model anymore), there is always room to attack the aptness of the analogy or comparison.

Any of these warrants can be sound or unsound. Definitions can be legitimate or perverse; samples can be representative or biased; effects can be good clues or just circumstantial evidence; constructs can be relevant or irrelevant; models can be competent or inept.

For any of these warrants, your proof can be sound and convincing or (as shown in Table 5-3) fallacious and specious.

TABLE 5-3 How Warrants are used in Both Sound and Unsound Arguments

WARRANT	SOUND PROOF	UNSOUND PROOF
Definition	Deduction; proof from standards	Begging the question; using an unsupportable definition or standard; proof by name-calling
Sample	Induction, sample; probabilistic inference	False generalization; biased samples; unrepresentative data
Effects	Solving the riddle; detection, investigation	Circumstantial evidence; "sign reasoning"
Construct	Correlation; causation; functional interdependence; concomitant variation	Post hoc fallacy; effects confused with causes
Comparison	Case study; analogy; model; simulation	Inapt analogy; reductionism, oversimplification

Persuasive reports will rise or fall on their warrants. Except for the rare cases where the facts are off, persuasive messages succeed when the warrants are convincing. When there is a controversy, typically it is over the warrant, not the evidence.

Thus, if your warrant is strong, bring it out into your discussion. Explain your reasoning; give evidence or proof in support of the warrant itself. (Careful: You will introduce a new warrant to do that.) Challenge your adversary to dispute your warrant. Indeed, attack the weaker warrant of the other side's case.

If, however, your warrant is not so strong, you may feature it less prominently or even keep it as an unexpressed assumption. (Of course, if you believe your warrant is invalid or false, you should not be using it at all!)

How to Make a Report Motivate

Motivation, remember, is different from persuasion. Persuading, getting people to share your attitudes, opinions, and beliefs—as hard as it is to do—is still easier than motivating, getting people to act or move.

To move readers, you must overcome their inertia. In that sense, motivational writing must be dynamic: It must compensate for the reader's momentum.

Put simply, to get readers to act, to do something they might otherwise not have done, you must find out what they want most and help them to get it, or find out what they want least and help them to avoid it. Thus, an effective motivational communication convinces your readers that if they do what you advocate or recommend, then, they will either get more of what they value most (an *inducement*) or avoid losing what they value most (a *sanction*). If the readers believe your prediction, they will be moved to act.

To be a successful motivator, then—and all selling is included under this heading—you must know your readers' values and offer to gratify them. You must, in short, study your readers as though they were a market for your ideas and you must present your ideas in a way that is acceptable to the market.

Usually, when I talk like this about motivation, a great weariness comes over my engineering or technical audience. The word *motivation* conjures up images of psychology and management and other "soft" disciplines. It makes the practical, down-to-earth working man or woman think of Madison Avenue and exotic advertising techniques. This is a silly and useless prejudice that, incidentally, ruins many technical proposals.

Understand this: If you will not learn to motivate your readers, then you will probably not get very far in business, industry, or government. If you will not decide what you want your readers to do, and motivate them to do it with your messages, you will have great difficulty as a leader or manager.

The strategy of motivation is simple, not exotic. The motivational report begins with a statement of need or mission and ends with a recommendation or proposal. Some typical need or mission statements follow.

> We must find a less-expensive way to transport steel to our factory.

or

> We must find a way to get the agency to speed up its approval of our new analgesic.

or

> We are looking for ways to increase our market penetration in the sales of ascorbic acid.

or

> We need a way to justify our space exploration budget to the Congress.

or

> We must get the sponsor to award us an additional $10,000 for Phase VIII of the project.

The strategy of motivation emerges in the way you state and expand the statement of need or mission. Whether the reader will "buy" your recommendations at the end depends very largely on how you set up the need statement at the beginning.

In effect, the way you write about the need or mission determines what actions or proposals will be best. You "wire in" the selection criteria into the statement of need, ensuring

- First, that the criteria reflect the main values and motivations of your reader; and
- Second, that the action that follows most directly from these criteria is the one you favor.

There is no dishonesty in this. There are hundreds of legitimate ways to come up with solutions and actions that will address needs and achieve missions. But there is no necessary reason for you to *present* your recommendations using the same logic that you used to *derive* them.

If you know what motivates your readers, and you can show convincingly that what you propose gives the readers what they want, then do so. And to make matters easier, use the classification scheme below.

This scheme, adapted from decades of work by the Yale sociologist Harold Lasswell, describes what people want from their lives and their work. It provides a vocabulary for talking simply and understandably about what motivates people and how different people are motivated differently. In Lasswell's plan there are only eight motivators or "values" (presented here in no particular sequence or priority):

- **Power** — control over decisions, yours and others'
- **Wealth** — money, goods, services
- **Respect** — being thought better than other people
- **Affection** — being loved, belonging
- **Well-being** — being free of mental and physical discomfort
- **Rectitude** — doing the moral, "right" thing
- **Skill** — developing talents, being attractive
- **Enlightenment** — understanding how all values are related

Ever since 1965, when I first studied this scheme, I have used it successfully in my writing and consulting. Never have I wished that there were an additional value on the list; never have I found that I could do without one.

Put simply: This scheme can be used to describe what your readers really want. As you can see in Table 5-4, each value can be used as inducement (get some more of this) or as a sanction (avoid losing this).

Use inducements more often than sanctions. When people trust you (not absolutely, of course, or else there would be no need to motivate them), they will

Table 5-4 Inducements and Sanctions for Motivational Communications

APPEAL	INDUCEMENT (Get These)	SANCTION (Avoid These)
Power	Make more decisions; Have more control; Be more independent, free.	Make fewer decisions; Lose control to others Become dependent, vulnerable.
Wealth	Have more money; Have more goods and services.	Lose money; Lose goods and services.
Respect	Win admiration; Be regarded as best.	Lose admiration; Be thought of as also-ran, loser, number 2.
Affection	Win friendship Be regarded with warmth; Be one of the group.	Make enemies; Be regarded hostile; Be an outsider, outcast.
Skill	Get know-how; Advance the state-of-the-art; Exploit talents.	Lose knowledge; Use second-best methods; Neglect exploitable resources.
Rectitude	Do the right thing; Be moral; Salve your conscience.	Break the law; Be immoral; Be guilty.
Well-being	Be healthier; Feel better; Live longer; Grow strong.	Get sick; Feel lousy; Die; Muddle along.
Enlightenment	Understand life; Know your own mind.	Be confused; Be neurotic.

respond best to inducements. If you threaten them with sanctions, they may believe you, but they will also tend to think less well of you.

In contrast, when you are writing for readers who dislike and distrust you, use sanctions. To hostile readers your inducements will seem dishonest, like intellectual bribes.

If you are not sure whether your audience is trusting or hostile, you will do better to err on the side of trust and use inducements.

6
Designing Proposals

GOOD PROPOSALS: THE CRAFT OF WRITING TO WIN

Engineers or scientists who can write convincing technical proposals will earn more money and get more satisfaction from their work than those who cannot. Few predictions are safer than this one.

A technical proposal is the instrument with which organizations and individuals find, win, and hold the money they need to survive and grow. And the higher you rise in the ranks of American technological industry, the more likely you are to be a writer or manager of technical proposals.

Put simply, an effective proposal may be defined as

> An assertive statement, in which an offeror convinces a sponsor or customer to purchase some product, service, or project.

Obviously, then, a proposal is a motivational message intended to get a reader, called the *sponsor* or *customer,* to spend money or lend other material support to the writer, called the *offeror.* A good proposal must *sell;* it must use appeals and motivators that cause the readers to part with some of their money. A document that does not sell—does not argue, cajole, convince, or move—is *not* a proposal. And yet, as obvious and basic as this claim may seem, too many engineers and scientists write proposals as though they were casual informative reports, as though a simple recitation of information could move a sponsor to spend thousands or millions of dollars!

For a proposal to be effective, it must prove four claims, none of which is a simple matter of information, and all of which require "selling":

First, it must convince the sponsor that the offeror has just what the sponsor wants and needs—not approximately, but *exactly.*

Second, it must give the sponsor complete confidence that the offeror can do what is promised, that is, can execute the plan in the proposal.

Third, it must state and justify a price that seems fair to the sponsor, or, if possible, an excellent value.

Fourth, it must demonstrate that the offeror's ideas, reliability, and price are more desirable than those of the competition (if there is any).

Further, proposals must be written for readers who are under stress—as are all customers about to make an expensive purchase. Especially when the proposal is in response to a competition, when there are three or four (or even ten or twenty) firms chasing the same contract, the reader is almost frightened to select one from the many. Usually, when a proposal is approved for award (especially if the government is the sponsor), there follows a chorus of complaints and criticism. The losing offerors, as well as their friends and lobbyists, are likely to grouse about the choice.

Proposal readers, therefore, are extremely close and careful critics (for the most part). And the more competition there is, the more the readers will claim to find fault in the proposals. Faults, flaws, technicalities—these reduce the number of competitors, making the choice easier and safer, reducing the burden on the harassed proposal reader.

Good proposals, therefore, must be technically excellent, understandable and convincing to a skeptical or reluctant reader, and completely free from any of the hundreds of potential flaws that would allow a reviewer to toss the proposal into the outbasket. Even under the most relaxed of circumstances, proposals would be hard to write well.

But the circumstances under which proposals are prepared are far from relaxed. Quite the contrary, they are written with insufficient time, on evenings and weekends, occasionally by teams of people who have trouble getting along. Typically, they are written amid conflicts between the marketing and the technical interests of the firm, and often with a sense of impending doom should the proposal fail to win. Further, the deadline for proposals is usually so short that there is no time to "remake your hand," as they say in rummy. The strategy and outline you prepare in the first hours or days will have to sustain the final product.

Some firms schedule a last-minute "red team" review of the proposal at the end of the process. The "red team," senior technical and marketing experts, have the power to demand major revisions from the authors. Consequently, the last revision of a technical proposal is sometimes done by people who have not slept for days.

For these and other reasons, proposals require a more careful plan and design than any other document you are likely to write. Major revisions should be done in the "storyboard" or outline stage. Although badly planned reports can cause a loss of time and money, badly planned proposals can cause a loss of time, money, and health!

DEVELOPING A STRATEGY AND THEME

Most of the effort in planning and preparing a proposal is spent on making sure that it is complete; that it meets all the specifications, provisions, and stipulations; that it observes all the rules and technicalities. This is the *compliance* side of proposal writing, the part in which you make sure that your proposal will not be thrown out by some low-level bureaucrat with a checklist.

If your proposal is free from flaws, it will be read. Freedom from flaws, however, is rarely enough to win. A winning proposal must have a positive impact (as well as a lack of defects). It needs a theme, an image, an identity. Even before the outline can be written, the author (or manager) of the proposal must be able to answer one or more basic questions: *Why me?* Why should this sponsor award this contract or grant to me and my organization? What special virtues or insights do we have? What right do we have to offer our services? What advantages do we have over the competition?

These questions are usually asked retrospectively. Most proposals are sold as much on the basis of what the offeror has done up until now as on what the offeror proposes to do. Further, the typical goal of the proposal writer is to generate work and sales for the existing staff, products, or systems—not for new ones acquired especially for the current offering.

The central strategic question, then, in designing a proposal is

> How can I present my already existing personnel, equipment, facilities, and other resources so that they seem tailor-made for the needs of the sponsor?

The answer to this question is your "win strategy," as some proposal writers call it. Usually it comprises one or more *themes,* recurring statements that tie your proposal together and justify the merits of your approach.

The point to be stressed here is that a proposal strategy is developed neither forward (that is, by building a proposal that exactly meets the requirements of the sponsor) nor backward (that is, by building a proposal that merely uses whatever resources you currently have), but rather in a loop. (See Figure 6-1.)

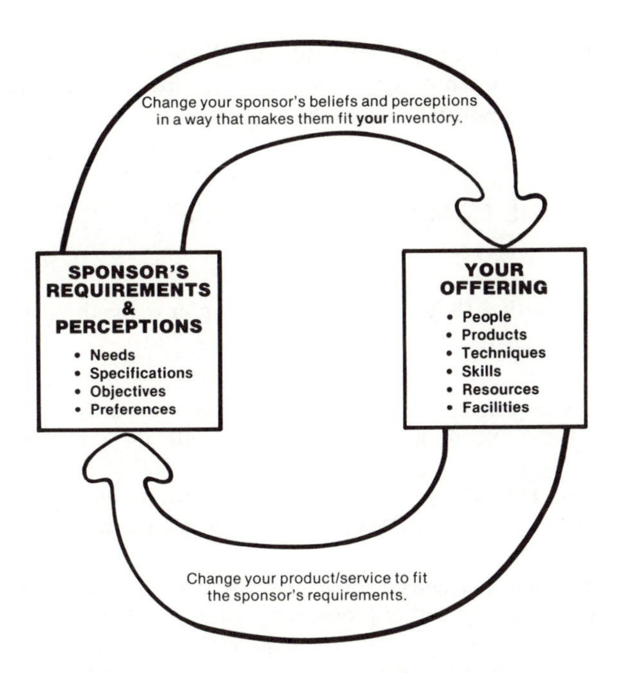

FIGURE 6-1 The Proposal Planning Loop

Although there will occasionally be a perfect fit between what the sponsor wants and what you already have, the typical case is quite different. Typically, you begin by trying to convince yourself that, effectively presented, your existing resources can be made extremely attractive to the sponsor. If that is not possible, you explore ways of changing your existing resources—adding staff, for example—to better meet the sponsor's demands. (Unfortunately, sponsors are more impressed with resources you already have than with resources you promise to acquire.) Each time you change your existing package of people and resources you ask again, Can we now make a winning offer to this sponsor?

If, at some point, you decide that you cannot convince this sponsor that you have whatever is necessary—at least not as well as the competition can—you may then decide to stop writing the proposal. Stopping it now is BETTER than waiting until you have spent more money.

Once you have developed your strategy, once you can write three or four strong thematic sentences that explain why yours is the most worthy offer, you are ready to outline.

Because the most useful way to write a proposal is to do your major revisions during the *planning* stage, not during last-minute repairs, you will want to make a detailed outline: the more authors, the more detailed. Most major defense and aerospace contractors plan their proposals in the storyboard or modular form discussed in Chapter 4, making sure that the headlines of the modules correspond to the specific Scope of Work requirements in the RFP. When you can identify the module or section of the proposal in which every single requirement or theme will be discussed, you are ready to write the proposal.

THE COMPONENTS OF AN EFFECTIVE PROPOSAL

Most proposals, whether they are three pages long or three thousand, contain the same three components:

The Technical Component, including a discussion of the background or problem; a description of the mission and goals; and a presentation of the methods or approach

The Management Component, showing the work plan or schedule; the personnel and organization; management control methods; corporate capabilities and experience

The Budget/Business Component, including the cost, cost breakdowns, and justifications; assurances and certifications; accounting reports or financial data

In addition to these three essential components, there are two subsidiary components:

The Front Matter, including the frontispiece, title page, index, contents, lists of exhibits, acknowledgements, abstract, and summary

Attachments, including brochures, resumés, testimonials, papers, and miscellaneous appendices

Keys to the Technical Component

The goal of the Technical Component is to persuade the sponsor (or the reviewers) that you know *exactly* what the sponsor needs and that you are prepared to do *exactly* that. Thus, the Technical Component contains most of the strategy you have developed in the planning stages. And, usually, all of the engineering and design work is written up in the Technical Component.

The Technical Component is a three-step analysis:

- Given the sponsor's real need (Introduction),
- Certain products or services will be offered (Offering),
- In this particularly appropriate way (Approach).

The details of this analysis can be seen in Figure 6-2.

Typically the opening of the Technical Component (usually called the Introduction, Background, or Discussion of the Problem) begins with a "plug" for the offeror. For example:

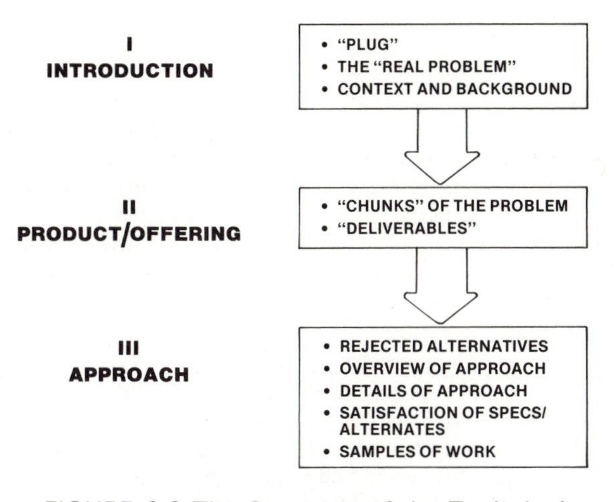

FIGURE 6-2 The Structure of the Technical Component

> OMEGA Company, developer of the INFON data-base management system, is pleased to present . . .

or

> ABCO, Inc., which has provided engineering consultation to Hickstown for the past twenty years, is pleased to . . .

Because reviewers may read the same words differently, depending upon the source of the words, tell the readers who you are—right away—and let them know that they are dealing with someone important.

Once the "plug" is out of the way, tell your sponsor the "real problem." This is the moment when you assert your theme and begin to reveal your particular approach. Knowing what you intend to offer, and knowing (from your research) what will most motivate your sponsor, you take a deep breath and state what the sponsor *really* needs. To do this, you must go beyond the information in the solicitation, beyond the published documents; you must say something special and insightful: push the sponsor's "hot button."

This is the riskiest moment in the proposal (with the possible exception of the price line). You judge what will most impress the sponsor, preferably some idea or analysis that will make you appear insightful *and* more aware than your competitors (if there are any). If you guess right, you are off to an excellent start; if you guess wrong, you are in serious trouble.

Over the years I have developed a kind of formula for stating the real problem. In one paragraph, I put

- A Technical Statement of the problem
- A Catch Phrase restatement of the problem, underscored
- A statement of the main motivational theme

Here are two examples:

1. **NASA Project**
   The aim of this project is to survey those city and state problems that lend themselves to technical solutions and search for items of recently developed aerospace technology that can be applied to those problems. *(Technical Statement)* In other words, *to put space technology at work on earth. (Catch Phrase)* In this way, NASA will be better able to explain the benefits of aerospace research to those critics who doubt the relevance of the NASA program in times of economic austerity. *(Theme)*

2. **Public Transit**
   The goal of this project is to design a high-speed rail system for Harbor City in which the vehicles and service will be so comfortable and attractive that commuters will prefer them to their private automobiles. (Technical Statement) In other words, *a public transit system that people would use even if they didn't have to. (Catch Phrase)* Only through the offering of greater comfort, safety, and convenience will Harbor City ever induce its commuters to prefer public transportation to increasingly undesirable private transportation. *(Theme)*

The catch phrase is more important in large government-sponsored projects or in any project (governmental or corporate) where a lot of people, including the press, will be writing and talking about the project. The "real problem" paragraph is a success when the sponsor reads it and concludes: I could not have put it better myself!

After the statement of the problem, you should write about the *context* and *background.* The aim of this section is simple: prove that you know a lot about the sponsor, that you are an "insider." Discuss the recent history of the sponsor's agency or company (or even distant history if that is relevant); demonstrate your knowledge of earlier projects—especially the ones that were, in the sponsor's opinion, the most important, or the biggest successes or biggest failures. Show that you have even read reports that you were not supposed to have read (handling the matter delicately), that you have talked to key people, and that you know much more about the sponsor and the sponsor's needs than the average person knows.

Also, as part of the context, show your familiarity with the sponsor's plans for the future. Discuss how this project fits into the sponsor's larger aims or mission. In sum, do everything you can to assure the sponsor that your interest in the project is serious and enduring, that your proposal is not just an opportunistic grab for the sponsor's money.

The next element in the Technical Component is the *Product* or *Offering* itself: the goods or services you intend to deliver. Some writers put this at the end of the Introduction—as the culmination of the Background discussion—while others put it at the beginning of the Approach section.

The Product section has two parts: a partitioning of the "real problem" into "*chunks*" or components, and the offering of a particular product or "*deliverable*" to satisfy each "chunk" of the problem.

There are two approaches to this section, depending on whether the sponsor has already told you precisely what deliverable products and service are needed or whether the sponsor has left that up to you. In the latter case, you can "chunk" the problem and define the deliverables in a way that anticipates your approach and satisfies your selling needs. That is, if the sponsor has not dictated the deliverables to you, you can make them up in such a way that to produce them would make your firm and its approach the inevitable choice.

Notice that the art of this discussion of deliverables is to break the problem apart in a way that most helps you, the offeror, while convincing the sponsor that it is the only reasonable way to partition the problem. In the following example, a single problem—testing the impact of an organization's antismoking campaign—is partitioned in two ways, the first favoring a firm that specializes in survey research, and the second favoring a firm that specializes in media communications:

1. **Survey Research Breakdown**

- Name the communities in which the campaign was used
- Draw samples of subjects in the target group
- Conduct telephone survey to gauge overall degree of recognition and retention of antismoking announcements
- Conduct in-depth personal interviews with subsample to assess impact of campaign on smoking behaviors

2. **Media Orientation**

- Identify radio and TV stations using the campaign materials
- Collect information on number and time of airings
- Collect information on viewer (listener) response
- Discuss reasons for using the materials more or less with media executives
- Poll media decision makers on ways to make the campaign more effective

If, on the other hand, the sponsor has dictated all the deliverables, then the task becomes an interesting puzzle in which you *prove to the sponsor that this selection of the deliverables is correct and ingenious.* In other words, you develop some partitioning of the real problem that will make the sponsor's list of deliverables the most sensible way to address the chunks of the problem!

In Figure 6-3, the real problem is to predict the impact of a new highway; the sponsor has determined the required reports and other deliverables. The proposal writer had to invent objectives (chunks of the problem) to give order to the deliverables and prove how right the sponsor is.

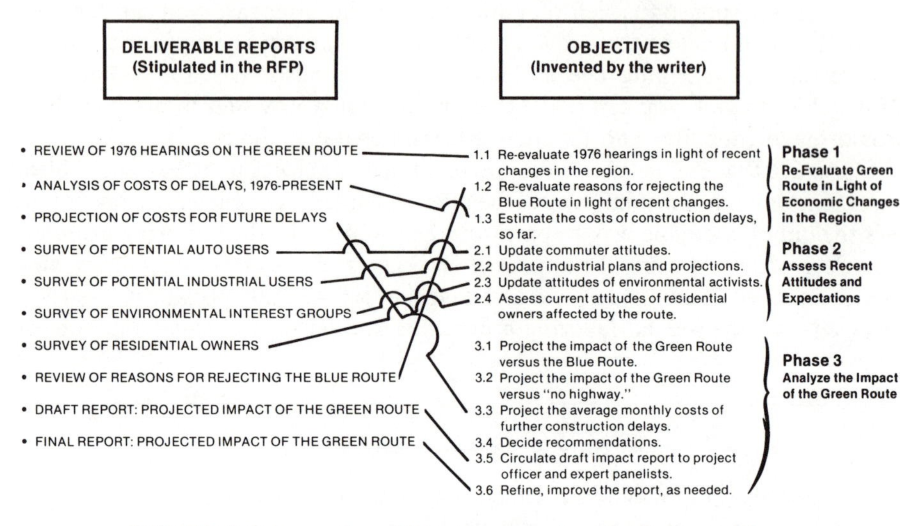
PROJECT: TO ASSESS THE ENVIRONMENTAL AND ECONOMIC IMPACT OF USING THE "GREEN ROUTE" FOR HIGHWAY 695
DELIVERABLE REPORTS (Stipulated in the RFP)
OBJECTIVES (Invented by the writer)
• REVIEW OF 1976 HEARINGS ON THE GREEN ROUTE
• ANALYSIS OF COSTS OF DELAYS, 1976-PRESENT
• PROJECTION OF COSTS FOR FUTURE DELAYS
• SURVEY OF POTENTIAL AUTO USERS
• SURVEY OF POTENTIAL INDUSTRIAL USERS
• SURVEY OF ENVIRONMENTAL INTEREST GROUPS
• SURVEY OF RESIDENTIAL OWNERS
• REVIEW OF REASONS FOR REJECTING THE BLUE ROUTE
• DRAFT REPORT: PROJECTED IMPACT OF THE GREEN ROUTE
• FINAL REPORT: PROJECTED IMPACT OF THE GREEN ROUTE
1.1 Re-evaluate 1976 hearings in light of recent changes in the region.
1.2 Re-evaluate reasons for rejecting the Blue Route in light of recent changes.
1.3 Estimate the costs of construction delays, so far.
2.1 Update commuter attitudes.
2.2 Update industrial plans and projections.
2.3 Update attitudes of environmental activists.
2.4 Assess current attitudes of residential owners affected by the route.
3.1 Project the impact of the Green Route versus the Blue Route.
3.2 Project the impact of the Green Route versus "no highway."
3.3 Project the average monthly costs of further construction delays.
3.4 Decide recommendations.
3.5 Circulate draft impact report to project officer and expert panelists.
3.6 Refine, improve the report, as needed.
Phase 1
Re-Evaluate Green Route in Light of Economic Changes in the Region
Phase 2
Assess Recent Attitudes and Expectations
Phase 3
Analyze the Impact of the Green Route

FIGURE 6-3 Inventing Objectives to Justify Deliverables

The *Approach* section should begin with a methodological discussion in which you *reject* certain approaches and methods. If possible, you should show that those approaches and methods most likely to be offered by your competitors will be unworkable. Of course, you never disparage your competitors; you merely prove that certain "obvious" or "popular" or "typical" approaches to this project (those you expect the others to propose) will be inefficient or ineffective or take too long or will yield unusable results. Your main argument should be that these rejected approaches will not get to the "real problem." (Remember, you set up the real problem so that you would be the best qualified to solve it.)

You should make your approach seem the only reasonable way to perform the project—but you should be cautious about the language you use to express this conclusion. There may well be among the reviewers people who favor one of the approaches you have rejected; if you inadvertently insult these people or suggest that their preferences are incompetent, you may stiffen their opposition to you. On the other hand, if you do not show any advantages to your approach, you may fail to win support.

Once you have made your own approach inevitable—and you should have been laying the groundwork for several pages now—you present the *overview* of your approach or method or program. *Never* begin the discussion of your approach with a detailed listing of tasks or activities. Rather, divide the project or program into five to seven main phases or parts. Ideally, devise a phase that corresponds to each "chunk" of the problem.

Again, your attack will depend on whether the sponsor has listed a detailed set of activities and tasks or left them up to you. If you are on your own, you may merely continue the detailing that began with your partitioning of the problem.

If, however, the sponsor has dictated tasks and activities, then, again, you have a puzzle to solve: to organize the sponsor's prescribed approach into a scheme that makes it *more* sensible and understandable, inventing an overview that works as a rationale (really a rationalization) for the sponsor's list of tasks and specifications.

No matter how the list is prepared, do *not* begin with the details; begin with a simple overview that puts the detailed tasks and activities into a coherent context. (If you are writing in the modular format, you may choose to write an overview module.)

After the overview, give the *details* of your approach. Be specific; go beyond the information in the specifications, but make sure that the reader can "check off" every specification on his or her list.

(In the case of extremely complicated projects or systems, you may want to have two or three levels of detail, with a "subsystem overview" at the beginning of each section. My own rule of thumb is that I never ask the reader to absorb more than eight or nine items in a single category. If I have ten or more items to describe, I will break them into smaller categories. If any of the categories has more than ten items, I will break it into smaller categories, and so on.)

I have found that one of the most useful ways to present a complicated project or system is as a series of "overlays," like the overlays used in overhead projector slides. In effect, I describe the project or system *completely,* at two, three, or four increasing levels of detail. If, for example, I were describing a proposed air quality monitoring program, I would probably give three consecutive descriptions:

- A simple, nontechnical overview for public administrators, politicians, and newspaper writers;
- An intermediate overview for environmental professionals and managers in the affected parts of government and industry; and
- A detailed description for the engineers and scientists who will have to judge the technical quality of my proposed system.

Remember that while you are describing your approach (and these sections are usually written by technical people), you must *prove your claims* and *explain your decisions.* Any time you state a fact that is not universally accepted, you must back up the claim. Any time you express an opinion—especially one based on your firsthand experience—you must build a case for that opinion. Any time you offer to do anything that deviates from the specifications and is not mentioned in the solicitation, you must explain *why* that is a good thing.

To the extent possible, you should meet every specification and requirement, and, to help the readers, *you should include a chart or table in which you yourself check off every specification and give the page numbers on which the spec is discussed.*

(Some people who call themselves publications engineers refer to this aspect of the proposal as "positive traceability," believe it or not. Whatever you call it, make sure every spec is covered and that the reader knows it.)

But suppose you cannot meet certain specs? What then?

The answer is simple. Always give sponsors what they want—first. Most of the time the reason you cannot meet a specification or requirement is that you do not *want to* (either for technical or for commercial reasons). The best practice is to meet the specification first and then offer your own alternates—*which you must then defend as superior to the sponsor's requirements while taking care not to insult the sponsor's judgment.*

Be sure the sponsor sees the cost or performance advantage of your alternate. In the following exhibit, the first version just *informs* the sponsor of a proposed alternate; the revised version *persuades* the sponsor of its superiority.

BEFORE:

PIANEX is 82 percent as effective as a sound insulator as SMUFFEL, which is called for in the specifications. In outer walls, PIANEX adds R-8 thermal insulation to the existing R-12; SMUFFEL adds R-2. The thermal insulation characteristics of PIANEX make it worthy of evaluation.

AFTER:

PIANEX, although slightly less effective than SMUFFEL as a sound insulator, is so superior as a thermal insulator that we propose to substitute PIANEX for SMUFFEL in all exterior walls. Granted, PIANEX is about 18 percent less effective as a sound inhibitor. But PIANEX adds a full R-8 thermal insulation factor to the current R-12, while a comparable quantity of SMUFFEL adds only R-2. So, given the urgency of current fuel conservation needs, the conspicuous advantage of PIANEX as a thermal insulator far outweighs its slight disadvantage as a sound insulator.

The hardest Approach sections to write are those in which the sponsor has already detailed everything that you should do, leaving you little to write. In some cases, extremely detailed specifications mean either that the proposal has been "wired" for someone who can meet those specs or that the quality of your proposal will count much less than the price.

When you cannot find anything to write about, then *give samples of the work.* That is, start doing the project in the proposal itself. If one of the tasks is to design a questionnaire, draft a tentative questionnaire. If one of the tasks is to review the literature on a certain problem, submit a tentative bibliography. If one of the tasks is to evaluate the adequacy of a particular plant, then write the criteria for the evaluation and make a tentative appraisal.

Actually doing some of the work can be difficult and risky. Even though you label everything as tentative or hypothetical, the reviewers may just plain dislike these samples of your approach. That would be most unfortunate if you had been doing well up till then.

Usually, though, samples of the work itself are the most dramatic way to prove to the sponsor that you know what you are doing—especially when you are a newcomer. I urge you to take the risk.

Keys to the Management Component

The goal of the Technical Component is to show that you *can do what you have promised.*

The Management Component should be an impressive display of your management skills, a proof of your experience and competence. To this end, you should include

- A Work Plan
- A description of your Management Controls
- A description of your Staff and Table of Organization
- A description of your Company's Capabilities and Experience

The *Work Plan* (jargon for "schedule") should contain, at the very least, a list of your activities and deliverables and their expected completion or submission dates.

A better approach is to include a Gannt chart (known familiarly as a "bar chart") showing either the duration of the activities or the key milestone dates. In Figure 6-4, Chart 1 is a traditional Gannt chart and Chart 2 is a Milestone chart. The most impressive work plan is a network (or precedence) chart, known popularly as PERT or CPM (Critical Path Method) charts. Although network schedules are a sophisticated management-planning device rather than a simple schedule, most proposal reviewers regard them as sophisticated kinds of work plans, proof of your technical competence.

(I sometimes feel that proposal reviewers are smitten with the artistic impact of a densely drawn network. It always makes your proposal more effective, and if you have not learned to draw these schedules, perhaps you should take the time to learn.) Figure 6-5 shows the three most popular formats for presenting network schedules. These specimens are quite simple, however, and it will probably take computer assistance to prepare the network for a project with more than twenty-five activities or tasks.

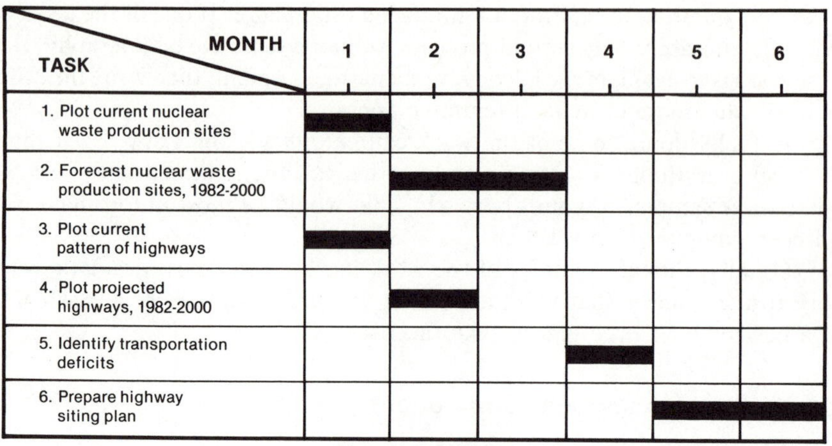

Chart 1 Activities

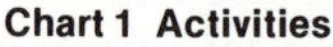

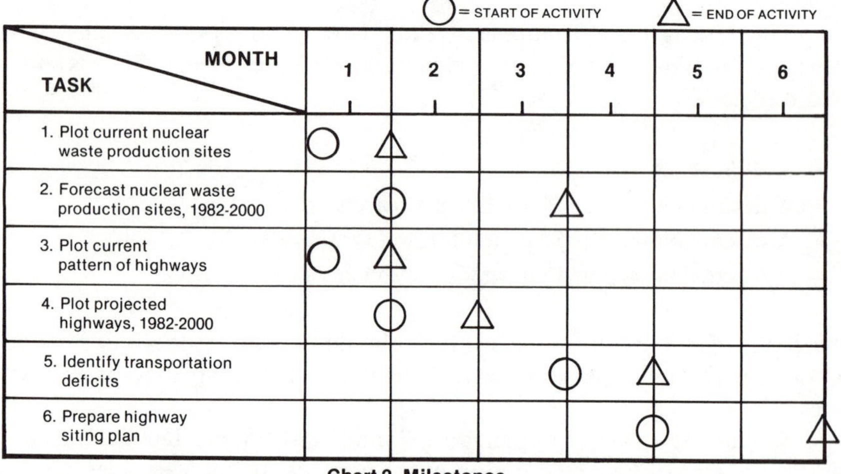

Chart 2 Milestones

FIGURE 6-4 Showing Activities in the Gannt Chart

Most Technical Components will also contain a discussion of your *Management Controls,* all the techniques your company will use to ensure high technical quality (performance), adherence to the schedule (time), and containment of expenses (cost). Your Controls should also include techniques that permit early detection of problems.

Also, you should describe the plan of meetings and reports you will use. Not just the "deliverable" reports called for in the RFP but also the regular, periodic progress reports—including not only those that you will send to the sponsor but those that you will collect internally from your staff.

Do not neglect the obvious. Describe your company's accounting system and give samples of the monthly financial printouts. Give the agenda for a typical

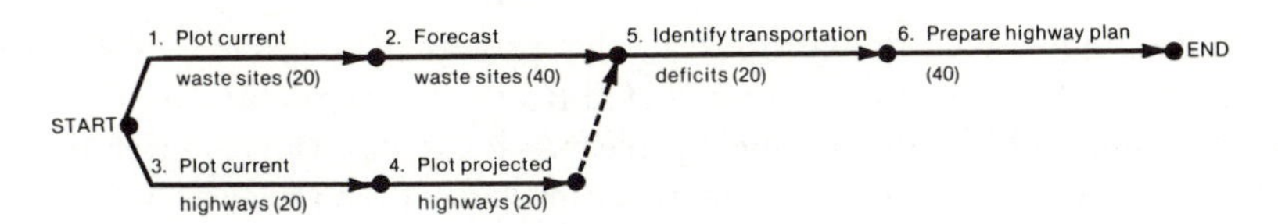

Version 1: Simple activity network

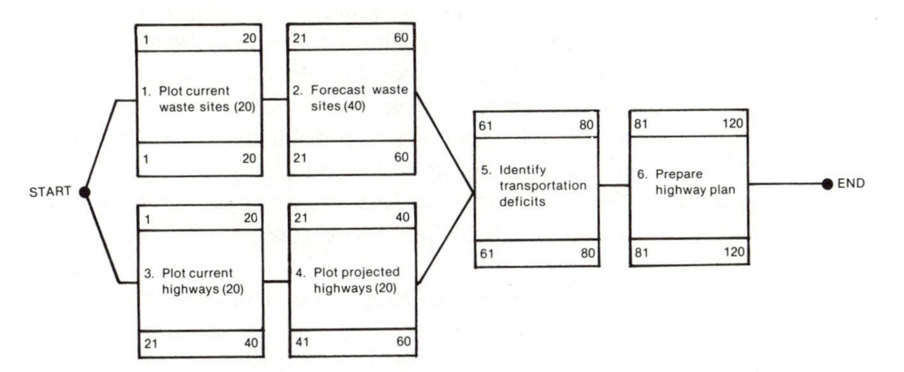

Version 2: Precedence network, showing earliest and latest allowable starts and stops

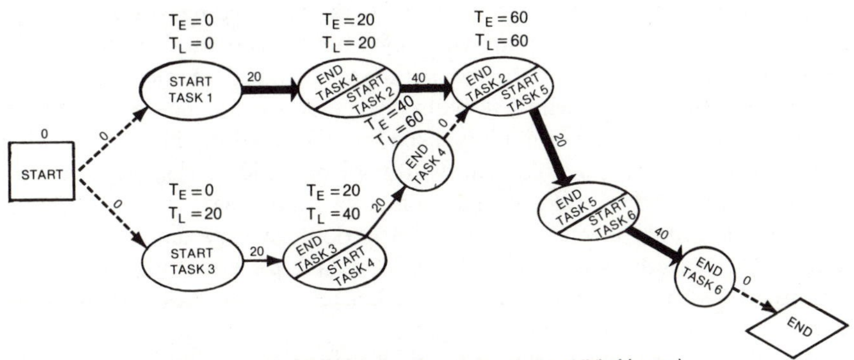

Version 3: PERT/CPM Network, showing "critical path" (bold arrow)

FIGURE 6-5 Three Ways to Show Activity Networks

monthly staff meeting. Do anything to assure the sponsor that you run a tight ship and will take good care of the sponsor's money.

If your work plan is in the form of a network schedule, then you have the option to present a whole series of automated project management reports. Today's network software (and there are scores of packages and systems to choose from) will produce any array of milestone reports, status reports, resource leveling projections, critical path analyses, line-of-balance analyses, estimates of the probability of finishing the project by various dates, lists of tasks sorted by any of a dozen descriptor categories. The opportunities are endless.

Interestingly, even though most of this project management software has been developed for tremendous, complicated R&D or construction projects—and is therefore too sophisticated for most projects—it can dazzle a reviewer and give him or her great respect for a new offeror, especially if the reviewer has not seen much of this technology before.

To make your presentation more effective, give sample printouts of the project control reports *with the sponsor's name appearing in the title.* Even though the data in the sample reports are fictitious and meaningless, the appearance of this name will make the sponsor feel that you are already on the job.

A caution: People who work with project management technology know that its principal benefit is in enabling the earliest possible detection of scheduling conflicts, delays, and overruns, and also providing information that will help the manager solve or ameliorate those problems. Even so, when you write about your controls, be careful of the words "delay" and "overrun"; most sponsors would like to believe that there will not be any.

Except in projects where you are offering equipment only, you will also want to describe the staff and organization that will do the work. (Even in "hardware" projects, I find it useful to name a project director or coordinator and any other people who will be accessible to the sponsor during the acquisition and installation of the equipment.)

Include a Table of Organization in which you *name names.* Few things give the sponsor more confidence than knowing *exactly* who the key people on a project will be; few things make the sponsor more nervous than references to some unnamed "senior analyst" who will be in charge of a major phase of the work.

Also include, whether or not it is asked for, a detailed staffing chart in which you say, as precisely as possible, who will work on which tasks, for how much time (expressed as hours, days, or percentage of total time).

Again, the more specific, the better. And do not worry about how difficult it is to predict exactly who will work on what task for how long. No one can predict that! Regard this as the science fiction part of the proposal, subject to continuous revision as the project progresses.

Specificity is the main virtue in the Staff section, especially nowadays when reviewers have become quite wary of firms that mention the reputations of their key people but do not specifically mention how much work they will do on the contract or project.

This section should also contain a pithy two- or three-sentence description of the qualifications and responsibilities of every key person in the project. Whether or not you intend to attach a complete resumé for each principal, you should also include a short, one-paragraph explanation of how qualified each one is.

Further, if the solicitation included a list of specific requirements for personnel, you should prepare a checkoff list (just as in the technical specifications) showing that you have included all the personnel skills required. (Better still, show that you have exceeded them.)

TASK / PERSON	PROJECT DIRECTOR (JONES)	SR. STATISTICIAN (SMITH)	JR. STATISTICIAN (BROWN)	PROGRAMMER (GREEN)	TOTAL
1. Plot nuclear waste production sites	2	3	5	5	15
2. Forecast nuclear waste production sites, 1982-2000	5	10	15	20	50
3. Plot current patterns of highways	1	1	10	5	17
4. Plot projected highways, 1982-2000	5	5	5	10	25
5. Identify transportation deficits	10	10	3	2	25
6. Prepare highway siting plan	10	10	10	10	40
TOTAL	33	39	48	52	172

FIGURE 6-6 Sample Staffing Chart (Expressed in Person-Days)

The Management Component usually ends with a discussion of your firm's *corporate capabilities* and *experience,* which may include a general description of your firm's history and structure and a detailed analysis of the firm's relevant capabilities and experience.

If appropriate, include concise summaries of earlier projects and activities.

If the proposal criteria state specific experience and capabilities requirements, prepare an exhibit demonstrating that your firm meets or exceeds those requirements.

In technology-intensive projects, the company's facilities and physical resources may be critical. Describe them in detail, pointing to specific items in the solicitation or proposal. Even in "less-physical" projects, the location of the operation will be relevant; emphasize, especially, the desirability of the location, the accessibility to the client and others, the transportation environment, or any other feature (such as an energy-conserving cooling system) that might endear the firm to the customer. (Note: Shabby, modest facilities should be sold as low-overhead. Remote, inaccessible locations should be characterized in terms of travel *times,* given the best possible connections.)

Other elements that may be important in this section are profiles of the project officers, a description of the company's finances and financial structure, unique company programs for employee health and welfare, social and community programs, and any other aspect of the company's philosophy or policy that bears on the current proposal.

Keys to the Budget Component

In most proposals for projects or services (as opposed to "hard goods"), the Technical and Management Components are reviewed independently of the cost and budget. In many cases, the cost data are required to be kept in separate documents and not discussed anywhere in the Technical Component (presumably so that technical reviewers will not be influenced by philistine issues of money).

There is not much real writing in the budget section. Usually it is just a series of tables, forms, and assurances, describing your price and backing it up. Sometimes, though, you believe that your price is higher than your competitors' or just more than your sponsor wants to pay. In that case, you need to persuade your reader that the price is justifiable.

Front Materials That May Help

Once you have attended to the "body" of the proposal, think about the front materials you will use to package it. In particular:

- **An Attractive Outer Cover or Frontispiece** shows that your firm is professional and successful.
- **The Title Page** should be clear and complete, referring to the solicitation if there is one.
- **The Table of Contents and Tables of Exhibits, Figures, etc.** should be complete, including every heading and item that appears in the proposal. Be sure to check for page numbers!
- **An Acknowledgments** page is useful, but only if you can cite people who will impress the reviewer; otherwise do not bother.
- **A Summary** is critical in commercial proposals. Ideally, it should be one page or less, containing a one- or two-sentence précis of each part of the proposal.
- **An Introduction** in nontechnical language belongs at the beginning of every technical proposal; in longer proposals you will need an introduction for each section or volume. The Introduction should explain all the parts that follow, give highlights of each part, and emphasize especially the selling points and themes. Think of it as a nontechnical abridgment of the entire proposal—the only part that the most senior staff of the sponsoring firm or agency will read. The Introduction is considerably more critical in the commercial proposal than in the

government proposal. But it is a good idea to add it to the government version, even if it is not asked for.

- **A Disclosure Protection (or Proprietary Statement),** citing appropriate laws or regulations, will either protect your proposal from abuse by the sponsor or give you a legal tool should you need to sue. In general, do not show proposals (or make presentations) to people you fear might steal your material.
- **Transmittal Letters** are usually unimportant formalities. If you want to discuss some important issue in the transmittal letter, though, then be sure to bind it into the proposal.

Whether or not a summary is required by the sponsor, you would do well to write one anyway.

A one-page summary (less than 250 words) is a succinct statement of your proposal strategy: why you believe that you have *exactly* what the sponsor needs.

The one-page summary, which would be the first text in the proposal, should have the following elements:

- A clear statement of who *you* are and a sentence that proves you are qualified to be submitting this proposal
- A one-sentence statement of the sponsor's "real problem"
- One or two sentences about the main components, objectives, or products you are proposing (keyed to the "real problem")
- One or two sentences about your method or approach, proving that it is the inevitable and correct approach to the real problem
- An overview of costs—*unless you are forbidden to mention prices in the technical proposal*

The style of the summary should be light, simple, clear, and fresh. No long-winded scientific or administrative jargon.

Attachments That May Help

Anything that is important to your proposal should have been in the proposal body. Most of the time, though, you will want to supplement the proposal with Attachments that may bring an extra rating point or two. Attachments such as

- **Complete, Long Resumés,** provided they are required or, if not required, very impressive
- **Long Brochures,** showing the full range of corporate capabilities
- **Long Project Descriptions,** including more-elaborate descriptions of methods and results
- **Excerpts/Samples** of earlier work, including documents, patents, drawings, construction (shown in photos or drawings)

- **Feature Articles** about the company, its operations, its key people, or its unusual managerial or social programs
- **Testimonials** from clients and sponsors of past projects or activities
- **Technical Appendices** that expand technical issues raised in the Technical Component of the project
- **Papers,** speeches, or reports by members of the project team
- **Stockholders' Reports** or other glossy accounts of the firm

Much of the material appearing in both your Management Component and your Attachments is referred to as "boilerplate."

"Boilerplate," as I told you earlier, is stuff that appears the same in many documents. Typically, it is from company brochures or data books or periodic reports. The best boilerplate is that which can be used "as is" in the proposal.

The most commonly used boilerplate is in the areas of company history and description, project summaries, employee summaries/resumés, facilities descriptions, policy and administrative procedures, organization charts, and so forth.

The principal advantage of boilerplate is that it is produced without any new work—or with almost no reworking. You merely have to assemble it and reproduce it.

Like all standardized things, though, boilerplate is rarely perfectly suited to a particular offering. Some parts are irrelevant, and others are counterproductive; and there are things that *should* be said but are not included in the standard texts.

The good proposal manager-writer will always want to fix the boilerplate to fit the current proposal. He or she will want to change every project summary, every resumé, every company description. But the demands on the manager-writer's time, and the pressures from the "new" parts of the proposal, will conspire to make him or her accept boilerplate "as is."

Struggle against that natural tendency. Polish up the boilerplate.

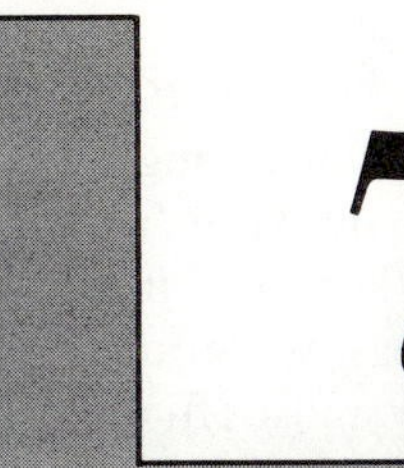

7

Designing Papers, Articles, Manuals, Memos, and Letters

PLANNING BEFORE THE DRAFT

So far we have talked about planning and outlining in general (Chapters 3 and 4), and reports and proposals in particular (Chapters 5 and 6). I hope that by now you get the idea: Everything you write must be regarded as a product whose function must be defined, and whose form should fit its function.

Most of the examples and illustrations have been for rather long messages, such as hefty reports and multivolume proposals. The principles, however, are the same for your shortest messages, even for your one-page letters and memos. Unless you have unlimited time, you *must* plan and design before you write your first draft. Otherwise your first draft will be your plan and, therefore, hardly usable.

The most important part of the planning—in my opinion, even more important than knowing one's audience—is defining the objective of the message, knowing exactly what you want the reader to know or believe or do as a direct consequence of reading the message. Further, once you know that objective, I submit that you also know where to begin and how to proceed. In short, I think that the general principles and specific applications you have seen so far give you a powerful planning technique that you can, and *should* apply to virtually everything you write from now on.

The aim of the rest of this chapter, though, is just to give you a few more illustrations and applications of the process, some further advice on the traditional design and structure of the messages most often written by engineers and scientists. Think of this chapter as pointers, rules of thumb to use when you begin to write your plan and outline.

DESIGNING A PAPER

An engineer friend of mine tells me that the difference between engineering and science is in their final products: Engineers make systems and devices; scientists make papers.

I don't know if that distinction is accurate, but I do know that there are large chunks of the engineering professions and even larger chunks of the scientific professions where the most important thing people make is publishable papers. The old "publish or perish" rule that we used to associate strictly with universities is now the rule in many corporate research centers as well.

Many senior scientists and engineers are under considerable pressure from their employers to publish superior papers in the most prestigious "refereed" journals. Such publication is very nearly a kind of elite advertising, with the firm not only underwriting *you* while you prepare the paper but also subsidizing the journal that publishes it. Indeed, this relationship so much resembles a form of advertising that the United States Postal Service, in the late 1970s, toyed with the idea of changing the postage rates for these so-called advertisement-free periodicals.

Before I describe the structure of scientific papers and tell you what editors look for, I must first discuss the prevailing confusion about the objective of the typical paper.

Almost every scientist I know (and many journal editors) believes that the purpose of a scientific paper is *to inform,* to communicate facts to other people in the scientific community. Most of the scientists who read the journals actually read *to be informed,* that is, to get facts. Not surprisingly, when I lecture to such a group of researchers about the importance of persuading and motivating, they usually nod patiently and then assure me that, although all that is very interesting, what *they* do is *inform.*

To my mind, the most important question about a scientific paper is whether it is, in fact, just informative or whether it is meant to be persuasive. In the "soft sciences" there is no argument; every paper in economics and sociology, for example, is meant to *persuade* you of the truth of the investigators' conclusions. But in the "hard sciences"—biology, chemistry, and to some extent biomedical research—I am not always sure.

This I do know: *The structure of a scientific paper is inherently persuasive.* It starts with a thought-provoking problem and ends with a conclusion (even though it is usually called a "discussion"). But in many cases it appears that neither the writer of the paper nor the reader is especially interested in anything other than the data in the middle of the paper. The problem is discussed only perfunctorily; the "discussion" section is dull and inconsequential; but the data are interesting.

I also know that even though many scientific papers end with a recommended action ("more research is needed"), most of their writers have no strong recommendations at all and don't really care if any further research is done. (If they do care, they will write another document called a proposal, rather than a paper.)

So, a scientific paper is a communication structured to persuade. If you choose that it be merely informative, then you must "go through the motions" of writing the persuasive parts. But when you mean it to be persuasive, you must apply all the principles of persuasion I have discussed so far.

The Parts of the Paper

The outline for the scientific paper varies from discipline to discipline, but most follow a design similar to that for the persuasive report:

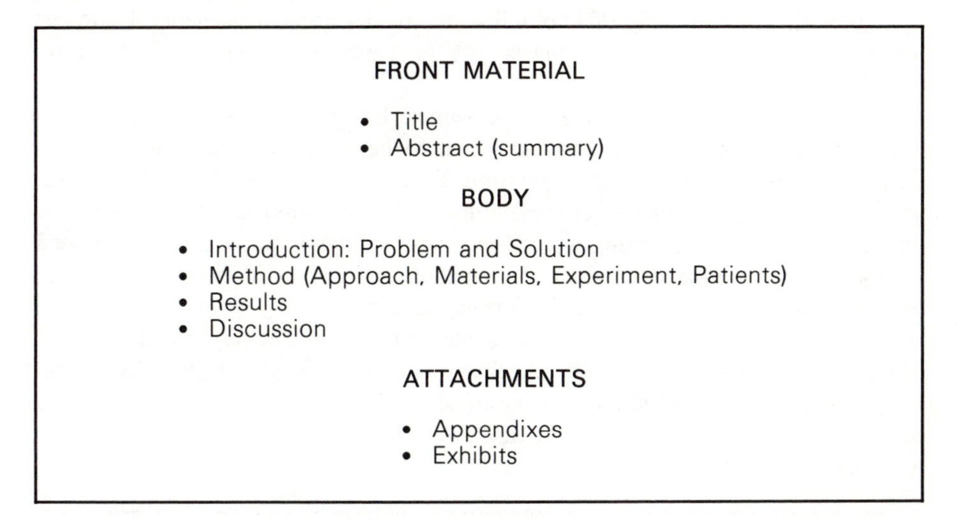
FRONT MATERIAL

- Title
- Abstract (summary)

BODY

- Introduction: Problem and Solution
- Method (Approach, Materials, Experiment, Patients)
- Results
- Discussion

ATTACHMENTS

- Appendixes
- Exhibits

If your paper has passed the review of a referee or a panel of experts, then the editors consider whether each of these parts of the paper is executed as effectively as it should be. (Some of the better referees will also give editorial advice, but most comment merely on the technical quality and suitability of the document.)

The title should be instructive. It must be complete enough to let potential readers know not just the subject but whether your perspective on the subject is relevant to them.

Instead of

METHIONINE TRANSPORT AND TUMOR DEVELOPMENT

write

INHIBITING TUMOR DEVELOPMENT THROUGH REGULATION OF METHIONINE TRANSPORT MECHANISMS

Instead of

STATISTICAL ASPECTS OF ATTENTION AND AROUSAL

write

HOW THE PROBABILITIES OF EVENTS ARE INVERSELY CORRELATED WITH ATTENTIVENESS TOWARD THOSE EVENTS

If these revisions look familiar, it is because they resemble the conversion from "headings" to "headlines," which I discussed in the chapter on outlining (Chapter 4). They have the effect of telling the reader not just the general category of the paper but also the emphasis, the problem, and even a bit about the approach.

The abstract should be a small summary. Long ago this statement was not true—the function of an abstract was just to help readers decide whether the entire paper was worth their reading time. Today, we try to perform that function in the *title* and let the abstract (one hundred to two hundred words, one paragraph) serve as a brief summary of the main findings and conclusions. Indeed, many abstracts are extracted from the paper and published in separate documents or distributed through technical information services.

As a consequence, most of the readers who read your paper will read *only* the abstract. For that reason, the old-fashioned kind of abstract, the "indicative" abstract that described what was included in the paper, has given way to the informative or summative abstract that presents a concise report of the findings. Consider these examples:

Indicative Abstract

A test of CA-701 in the treatment of subacute sclerosing myelitis is described. Problems in assessing the results are outlined, and brief descriptions of some of the side effects are provided. Further research on the effects of this drug in treating neurological disease is called for.

Informative Abstract

As part of a program to test the benefits of CA-701 in the treatment of several neurological diseases, we gave CA-701 by mouth to twenty-three patients with subacute sclerosing myelitis. Nausea was a dose-related side effect, with no patient able to tolerate more than 3 gm per day. After two weeks, nineteen patients showed fewer symptoms and none grew worse.

The first version does little more than mark the paper as potentially interesting to certain readers. The second version, though, is in itself a small, useful clinical report, valuable to many physicians and researchers.

The body of the paper begins with an *Introduction* that, at least, describes the problem investigated and, at best, gives a capsule version of the entire paper: problem, method, results, conclusions. Whether you confine your Introduction to the problem alone depends on the preferences of the journal you write for. But if your abstract has merely been indicative rather than informative, you had better turn your Introduction into a brief summary of the entire paper.

The best Introductions are "set pieces"; they are complete enough to stand alone, as a kind of longer summary. And they are short enough to need no subdivision or sections.

The *middle part of the body* is known by such names as "materials and methods," "patients and methods," "approach," "methodology," "experiment," and "experimental design." Usually the easiest part for most scientists to write, this section describes materials and technology used in the project, as well as procedures and techniques. Also, it includes enough references to other works (published tables, for example), so that a competent reader could duplicate the project or experiment.

Most editors prefer that you keep your results *out* of the Methods section, that you stick with what you did rather than what you found. Sometimes, though, when you are reporting a series of connected experiments, it is difficult to avoid mentioning how the results of Study A led you to the undertaking of Study B.

There are two parts left in the body of the paper; *Results* and *Discussion of Results.* Some editors will insist that you keep these two parts separate, that you first present the facts alone, without interpretation or embellishment. Other editors allow or encourage you to marry the two into a Results and Discussion section, permitting you to comment on your findings as you report them.

Whether the Results and the Discussion should be separated depends mainly on the ontological status of the conclusions. If they are uncomplicated, uncontroversial generalizations of the data (with confidence intervals attached), there is little harm in connecting them to the data. But if they are more speculative and interpretive—if the link between data and conclusions is less strong—you would do better to separate them, writing about them in a more speculative style than would be appropriate for the Results section.

Again, remember that you must decide whether your paper is informative or persuasive. If your objective is just to pass along the data, then your discussion need not be elaborate. If you want to prove a subtle point, however, and if you want to sustain a judgment that does not follow obviously from the data, then your "discussion" will have to employ all the techniques of persuasion mentioned earlier (Chapter 5).

Indeed, all important scientific papers are at least as much persuasive as they are factual. It is the very essence of science that investigators publish their *conclusions* along with their data and then wait to see if their conclusions will withstand the barrage of criticism and refutation that is sure to follow.

If you are writing about controversial matters, therefore, you will have to write your Discussion section as though it were one side in a debate. You will have to anticipate the objections that are sure to follow—even mentioning a few of them yourself. You will have to build a convincing argument—something that too few scientists bother to do.

Style in the Scientific Paper

Although I intend to talk about style later (Chapters 9, 10, and 11), I must say just a few words about the language and tone of the typical scientific paper.

Obviously, style does not matter much in the selection and publishing of scientific papers. I say "obviously" because the bulk of published scientific papers are in an execrable style. Apparently, many of the referees and editors who approve the pieces for publication are extremely tolerant—or perhaps do not know any better!

Writers of scientific papers can get away with more than other writers. Because the *content* of the work is so urgently interesting to the readers, the readers will put up with a denser and more difficult style.

Curiously, most scientists who write papers think that the dense, dull, stiff, inhibiting style of the scientific paper is necessary to create the appropriate air of objectivity and detachment. Nonsense! There is no reason to write

> It has been projected that there will probably be a significant incidence of allergic reaction to CA-701 evidenced by that subpopulation which is susceptible to penicillin.

when you can write

> Probably, many of the people allergic to penicillin will also be allergic to CA-701.

The second sentence is no less detached, impersonal, or professional than the first. It is just *better*. But more about style later.

DESIGNING A MAGAZINE ARTICLE

Writing articles for magazines is fun. It improves your writing; it wins you admiration from colleagues and family members; it can even earn you a few dollars.

I am not speaking now of the well-known magazines, like *New York* or *Harper's* or *Penthouse,* but of those thousands of lesser-known periodicals: trade and industrial magazines, hobby and avocation magazines, popular scientific and engineering magazines, house organs, employment chronicles, throwaways, Sunday supplements.

Although many of these magazines are slick and glossy products, written by staffs of professional writers, most of the magazines published in America are on slender budgets and depend on amateurs and free-lancers. At any given moment there are, literally, thousands of magazines in need of an article or two —and you probably write well enough to give them something they can use. (At most magazines, the editor will improve the style of your article, sometimes beyond a point where you can still recognize it.)

But how different are articles from papers?

Usually, articles are in a lighter, less-formal style than papers. But I've seen articles that were as intimidating as the worst paper. And besides, if papers were better written, they would be less imposing.

Usually, articles are shorter, less than twenty-five hundred words, but there are many scientific journals that publish papers shorter than that.

Usually, articles are freer in organization than papers and are less likely to be forced into a standard format or organization. But some magazines have rigid rules for the length and format of their articles, and some scholarly journals permit almost any organization for the papers they publish.

Usually, magazines are more interested in pleasing their readers than journals, but journals need the good will and respect of their readers too.

Articles differ in degree, rather than in kind. They are lighter, shorter, more personal, more entertaining, more relaxed, more friendly than papers—but not always and not necessarily. Perhaps the most consistent difference between them is the way they begin.

The Structure of Magazine Articles

Scientific papers and those articles published in serious technical magazines (*Science,* for example) begin directly and informatively. As I said earlier, the title should be instructive, the abstract informative. The readers should know at once what the piece is about and be able to decide whether the piece is worth their reading time.

Articles are less direct and more seductive. Although a paper begins with a straightforward statement of what the piece is about, without mystery or suspense, an article will often tease the reader. Articles may begin by setting a scene or creating a mood. An article can begin with a rush of action or with a slow, suspenseful buildup.

Magazine articles, written as they are for a less-committed reader, placed in a location where they compete for attention with expensive multicolored advertisements, must "hook" the reader. Certainly, sometimes a simple announcement of the subject will do the hooking. But more often, the writer will have to dangle a lure with the hook, to make sure that the reader bites down hard on the story. In sum, a well-written magazine article makes the reader *more curious about your subject than he or she was before.*

Let me illustrate. Suppose you were writing an article about a new kind of automobile that used a combination of diesel and electrical energy to achieve greater fuel economy.

If you were writing a scientific paper (or a very sincere technical article), your title would be straightforward and instructive, something like

> COUPLING DIESEL POWER WITH ELECTRICAL POWER FOR FUEL-EFFICIENT AUTOMOBILES

If you were writing a magazine piece, though, the title might be

> THE X-109 VERSUS OPEC

or

> MUSCLE CARS FOR TOMORROW'S KIDS

or

> EIGHTY MILES TO THE GALLON—AND FOUR WHEELS TO BOOT

The same thinking can be applied to the beginning of the article. The straightforward "scientific paper" way to begin might be

> Omega Engineering has developed an electrically powered automobile in which the batteries are charged by a small, diesel-powered generator.

But there are other ways to begin. Dozens of them. Here are a few hypothetical examples:

> It was 3:00 P.M. and Omega's X-109 had been looping the one-mile test track since noon . . . on one gallon of diesel fuel.

> When Omega Engineering's Tad Johnson was seventeen years old, his greatest pleasure in life was a 1932 Ford with a chrome-plated engine from a 1957 Oldsmobile. For Tad, the saddest part of today's energy crunch is that the next generation of kids may never know the sheer ecstatic pleasure of a big V-8.

> Omega Engineering's X-109 will get eighty-three miles per gallon on diesel fuel . . . as long as you don't turn on the radio.

> Burt Rogers is a security guard. He's 6′4″ tall, weights 240, and until last July was a guard in a stereo store in downtown Detroit. ("While I was on," he says proudly, "nobody lifted nothing.") Since July, Burt has had new duty. He makes sure nobody lifts, or even sees, Omega Engineering's X-109.

You get the idea. Before you announce the topic, before you explain just what Omega Engineering is up to, you make some personal, interesting contact with the reader. How much of this teasing and luring is appropriate depends on the style of the magazine. Some straight-laced periodicals will have none of it; others will let you build anticipation for three or four paragraphs. Eventually, of course, you will get to that straightforward announcement ("Omega Engineering has developed an electrically powered . . ."). But, in many ways, the most important part of the article is what you do before you get to the announcement.

Taken together, the lure and the announcement constitute the first part of the magazine article. (The announcement is referred to variously as the "thesis" or the "billboard.") It is the first of what are usually (but not necessarily) three parts of the article:

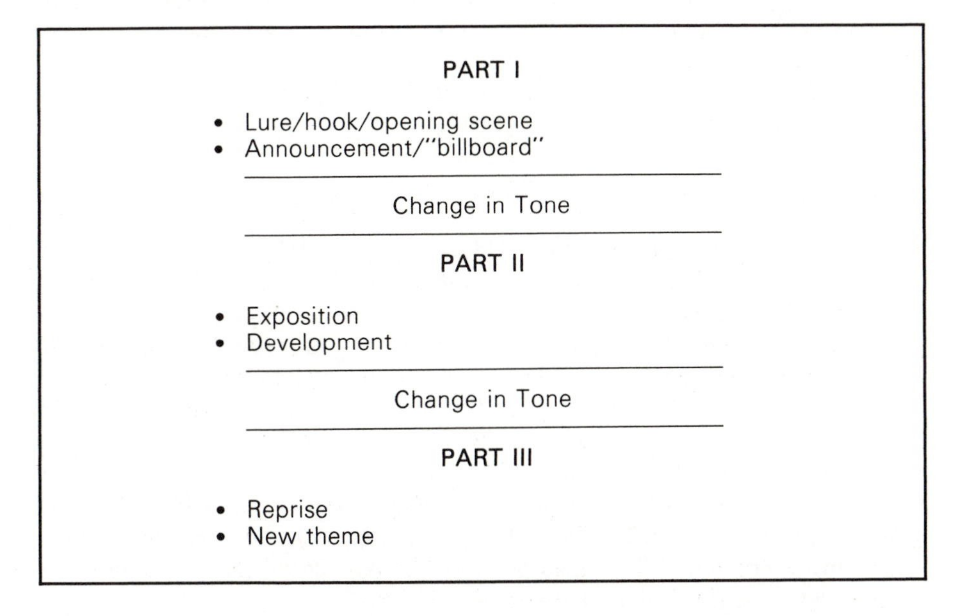

PART I
• Lure/hook/opening scene
• Announcement/"billboard"
Change in Tone
PART II
• Exposition
• Development
Change in Tone
PART III
• Reprise
• New theme

After the announcement there is a change in tone as you shift, typically, from being provocative or seductive to being informative and descriptive. The middle of the article contains your "research," what you have learned from thinking and reading about the subject. The most useful and interesting way to fill out the middle of an article is with *interviews* in which you quote people. (In technical papers, of course, you only quote other papers; in magazine articles there are actually real people speaking.)

The middle of an article does two jobs. *Exposition* is equivalent to "background": the events and facts you need to understand what is happening now. Notice that a good article will almost never begin with the background and exposition. It's too slow and dull; no one would take the bait.

After exposition comes *development,* talking about the side issues, the implications, the subtleties of the main story, the deeper meaning behind the events.

The middle part ends with another change in tone, usually from informative and descriptive to reflective. It's often a good idea to conjure up the original image again (see that car still looping the track or Burt Rogers still patrolling the plant), but there is no need to recapitulate what you have already covered. Sure, if a story has been complicated you may want to tie things together and summarize. But recapitulation is neither necessary nor particularly desirable at the end of a magazine article.

Most of the best articles end with a new theme, an idea or an image that spins off from the main story. You raise an interesting question, or predict the consequences of current developments, or speculate about what would happen if . . .

Never let the first or last sentence of your article be a question—that's too amateurish. But always leave some question or curiosity with the reader, some sense that there is still another story to tell.

Find a magazine you like to read—one written by top-quality writers—and see what I mean about the endings. You'll see how it is done.

Placing a Magazine Article

Let's assume that your head is full of ideas that you might write articles about, and that there are a few you would really enjoy thinking about and researching. How should you proceed?

First, pick the magazine where you would like your article to appear. Although some writers are so good that any magazine would publish their articles, most working writers have to adapt to the magazine. After you decide to go for a particular magazine, you have to study it, read several issues, notice how long the articles are, and decide what subjects and philosophies are most prevalent.

Second, after you have studied the magazine—and if you still like the one you have picked—*write away for that magazine's guide for authors.* Most magazines have them and will be glad to send one out. Study the guide until you understand what the editors demand of a writer.

Third, send a query letter. Write to the editor, briefly describing your idea, asking if the magazine would be interested. Query letters are useful in preventing you from writing an article that would be unacceptable—or from writing an article very similar to one that the magazine has already acquired. If the editor says yes to your inquiry, it is not a guarantee that your article will be published, but at least you know the project is worth trying.

Fourth, write the article, according to the Writing System. Define your purpose and think carefully about the needs of your readers—in this case the readers of that particular magazine. (If you're not sure who they are, find out.)

Fifth, send in the manuscript. Keep a log of the place and the date you sent it.

At this point, the process becomes more complicated. How long should you wait for a response? That depends on whether you are willing to wait. What should you do if the editor wants you to revise it? That depends on whether you can live with the changes the editor wants—or if you have the time or initiative to do the rewriting. What if you are rejected? That depends on how tough-minded you are.

Most articles can be resubmitted without much change to at least one or two other magazines. If that is true of your article, try again.

If none of the magazines you know about will accept it, go to the library and get the names of lesser-known periodicals and try them.

Be encouraged. There are so many magazines in America, with such an unyielding need for short interesting articles that, as long as your piece has any merit at all, and as long as you are not asking a lot of money for it, sooner or later you will almost certainly publish it.

DESIGNING A MANUAL

Most of the people who call themselves technical writers and editors spend nearly all of their working hours writing manuals. (And their close cousins the catalogs, parts lists, and specifications.)

Manuals may be the most challenging and irritating kinds of writing. They are certainly the most tedious and expensive. It is not uncommon to spend as many person-months writing manuals as building the system that the manuals describe.

Surely, I cannot tell you much about the detailed design of manuals in the next few pages. The U.S. military alone has tens of thousands of pages of guidelines and specifications for contractors who have to prepare manuals for military hardware. And every manufacturer of complicated equipment, from computer systems to water-ice machines, has its own ideas about how manuals should be written and what they should cover.

Manuals are hard to plan because manuals are *not read.* Manuals are *used* or *consulted* or *followed* or *studied.* Consequently, it is difficult to anticipate all the audiences for a manual and all the uses to which the manual will be put. Yet that is exactly what you must do before you write a manual or, more correctly, a *set of manuals,* since they almost always come in sets.

Functions and Users

Every manual (or set of manuals) must begin with an analysis of which aspects of the equipment or system must be written about (the *functions*), and of which people will need to have the information about those functions (the *users*).

Figure 7-1 illustrates the analysis that should precede the preparation of all manuals, namely, a listing of the relevant functions and relevant "users" for, let us suppose, a reasonably complicated piece of X-ray equipment. (These same categories of functions and users might apply to many other kinds of equipment as well, anything from industrial sewing machines to videotape recorders.)

The functions for this equipment are

USER / FUNCTION	MANAGER	BASIC OPERATOR	ADVANCED OPERATOR	MAINTENANCE TECHNICIAN
BACKGROUND & DEVELOPMENT				
OVERVIEW & GENERAL DESCRIPTION				
COMPONENTS & OPERATING PRINCIPLES				
STANDARD OPERATING PROTOCOLS				
SPECIAL APPLICATIONS/ ADVANCED FUNCTIONS				
FIELD MAINTENANCE				
MAJOR MAINTENANCE				
PARTS AND SPARES				

FIGURE 7-1 The User/Function Matrix

- **Background and Development** — the needs and objectives that led to the development of this machine, the improvements needed in existing technology
- **Overview and General Description** — what the system contains and what it can do, its several main parts and its basic abilities and services
- **Components and Operating Principles** — how the system works, the science and engineering that underlie its operations, the rationale behind its instrumentation and physical appearance, why it performs as it does
- **Standard Operating Protocols** — basic, step-by-step routines to follow in using the equipment for its most common functions
- **Special Applications/Advanced Functions** — more-difficult and unusual operations, involving judgment and interpretation rather than the following of standard routines
- **Field Maintenance** — regular maintenance to be performed according to a schedule
- **Major Maintenance** — ways to handle major breakdowns and subsystem failures
- **Parts and Spares** — names, numbers, prices; availability of all supplies, materials, replacement parts, and other items that may need replacement

Now, the users:

- **Manager** — the head of the department that owns and uses the system
- **Basic Operator** — a technician responsible for nonjudgmental, standard applications of the equipment
- **Advanced Operator** — a physician or engineer who may experiment with novel or advanced applications
- **Maintenance Technician** — a specialist trained in scheduled maintenance, trouble-shooting, and overhaul of the equipment

Even though each piece of hardware has its own functions and users, I'm sure you can see that most systems and devices have similar functions and similar categories of users. And the more complicated the system, the more refined the analysis must be: smaller categories, more precisely defined.

In planning a set of manuals, it is good, though not essential, if the two dimensions of the matrix are independent, that is, if there is not some particular function exclusively associated with one class of user. So it is best to use a different vocabulary to name the two dimensions. In the example above, calling one category of users "maintenance technicians" might obscure the fact that some of the maintenance on the system is done by the group called "basic operators."

But you cannot really know if the categories have been named properly or refined enough until you begin to use the matrix. In my own planning I have used the matrix in two ways. First, in simple systems, I merely mark each block *yes* or *no.*

As you can see in Figure 7-2, certain categories of functional information are of virtually no interest to certain classes of users. If I were using Figure 7-2 as my actual plan, for example, the basic operator's manual would include only standard operating protocols and field maintenance. (I'm assuming for the moment that I'm going to have a separate manual for each category of users; but that is not always the best plan, as I shall explain later.)

In more-complicated systems, with more-complicated groups of users, I sometimes use a more-complicated approach. Instead of *yes* or *no,* I'll use 0, 1, 2, or 3. The 0 means that the category of information is irrelevant to the user; then, 1, 2, and 3 show the need for increasingly complete levels of detail.

In Figure 7-3, the matrix really has a third dimension—level of detail—and each category of user is reviewed not only for the categories of needed information but also for the extent of detail needed in that category.

So, you see that a set of manuals is compiled and aggregated from bits and "modules" of information. Under Background and Development, for example, I may have three short discussions, the second an elaboration of the first, the third an even greater elaboration of the second. Any given manual in the set will contain either nothing from this category or just the first; or the first plus the second; or the first, second, and third together.

(The scheme above is the incremental approach. I might also have composed three alternative versions, each self-contained, each increasingly more

USER / FUNCTION	MANAGER	BASIC OPERATOR	ADVANCED OPERATOR	MAINTENANCE TECHNICIAN
BACKGROUND & DEVELOPMENT	N	N	Y	N
OVERVIEW & GENERAL DESCRIPTION	Y	N	Y	N
COMPONENTS & OPERATING PRINCIPLES	Y	N	Y	Y
STANDARD OPERATING PROTOCOLS	N	Y	Y	N
SPECIAL APPLICATIONS/ ADVANCED FUNCTIONS	N	N	Y	N
FIELD MAINTENANCE	N	Y	Y	Y
MAJOR MAINTENANCE	N	N	Y	Y
PARTS AND SPARES	N	N	N	Y

FIGURE 7-2 The Yes/No Matrix

sophisticated and detailed. In that case, the matrix would have called for *one* of the three.)

Looking at the matrix also suggests the way the materials should be organized into volumes. If two or three categories of users have similar needs, then there is no need to compile separate manuals for each user. Indeed, if I wanted, I could compose detailed (level 3) versions for each function and merely provide the users with a copy of the matrix, which they could then use as a guide through the set of manuals.

Or, if I wanted to, I could merely compile one immense manual, with all the modules carefully coded, and provide the matrix as an index. Or I might put that entire encyclopedic "book" into a computer data base, which the various users could get at just by calling for the matrix code number of the module they want to see.

The possibilities are many. In enormous electronics systems they become astronomical, overwhelming to that average citizen who first sees a ten-foot shelf of manuals. But at the base of every manual set—whether it is a fifteen-page pamphlet to accompany a bicycle or twenty thousand pages to accompany a nuclear submarine—is the function/user matrix. Without this simple analysis, the manual writer really does not know where to begin.

USER / FUNCTION	MANAGER	BASIC OPERATOR	ADVANCED OPERATOR	MAINTENANCE TECHNICIAN
BACKGROUND & DEVELOPMENT	1	1	2	0
OVERVIEW & GENERAL DESCRIPTION	3	1	2	1
COMPONENTS & OPERATING PRINCIPLES	2	1	3	2
STANDARD OPERATING PROTOCOLS	1	3	3	2
SPECIAL APPLICATIONS/ ADVANCED FUNCTIONS	1	1	3	1
FIELD MAINTENANCE	1	3	2	3
MAJOR MAINTENANCE	0	2	2	3
PARTS AND SPARES	0	1	2	3

FIGURE 7-3 The Detailed User/Function Matrix

Documenting Computer Programs

If I were asked to predict the most important development in technical communication in the rest of the twentieth century, my answer would relate to the solution to that most exasperating of problems: How to document a computer program or system. In other words, "software manuals."

If there is one thing that programmers and system users agree on it is that the typical manual is horrible. The people who write them—usually analysts who can't write and wish they didn't have to—admit that they don't know what they are doing.

From the users' side, I have heard blood-curdling accounts of major software purchases in which the documentation and manuals were so bad that the customers had to invest in having their own manuals written for the vendor's software!

Perhaps the most ambitious effort to resolve this problem, to define a standard way of organizing and writing software manuals, is the Federal Information Processing Standards (FIPS) branch of the U.S. National Bureau of Standards. Its main interest, apparently, is to standardize the function categories in the function/user matrix and define a particular kind of manual for each function. The ten FIPS categories follow.

- **Functional Requirements Document** — history, background, and development of the system
- **Data Requirements Document** — data needed for or used in the system
- **System/Subsystem Specifications** 4 — what the systems analyst needs to know
- **Program Specification** — what the programmer needs to know
- **Data Base Specification** — description of a data base
- **User's Manual** — nontechnical guide to the applications of the system
- **Operations Manual** — description of the software and the environment in which it operates
- **Program Maintenance Manual** — ways to maintain the programs and operating environment
- **Test Plan** — procedure for testing all aspects of the system
- **Test Analysis Report** — report on the test results, describing the readiness of the system

In this scheme, every potential audience or user (not just the people for whom the *User's Manual* is written) must find what he or she needs in this set of ten documents.

The FIPS design—although it may seem complicated to you—is actually simpler than many of the documentation standards that have been used by the military over the years. It is an extremely useful way to start the user/function matrix, although it will certainly need some adjustment from time to time.

The quest for a standard scheme of functions and users, however, is only part of the problem. The real weakness in most software documentation that I have seen, even in the best planned and organized sets, is that the small pieces of the manuals—the detailed explanations of the commands, the descriptions of the techniques for updating a data base, and even the procedures for logging-on and logging-off the system—are so clumsily written and hard to follow.

From the earliest days, programmers, analysts, and computer salespeople have been enjoined to think of the user, to be user-oriented, to concentrate on the uses of the system and not the system itself. But these injunctions have largely failed.

Ironically, even though computer systems have become increasingly easier for ordinary people to use, computer documents have become more and more arcane. Computer professionals seem no more eager to speak and write in plain English than ever. In fact, with the tremendous growth in the industry, it is now possible for computer people to go for weeks at a time without having to communicate with a single "civilian." And their writing is generally among the most cryptic and dense of all the technical occupations.

The solution? I believe that we will soon have to do to software documentation what the computer industry has itself already done to programming. We will have to "structure" it, "quantize" it, turn the craft of writing manuals into smaller and smaller standardized routines.

If we follow the logic of the function/user matrix even further, we can see that every page of every "module" can also be viewed as a module. And that there are only a certain number of kinds of page in a computer document. What we will have eventually is "structured documentation," writing done in small, manageable chunks, each self-contained, each clear, each linked to the rest of the document according to one of a small number of "standard moves." All the genius and analysis in the writing of these manuals will be used in planning and designing them. The actual writing will be a snap, easy enough for the least-skilled writer to execute.

And if we make the modules small enough and logical enough, some day we may even get the machines to write their own documentation!

Designing Memos and Letters

Memos and letters—correspondence—are not so much another kind of writing as another way to package your writing. You can put a report into a memo or put a proposal into a letter; you follow those planning techniques appropriate for reports or proposals, subject to the special constraints of letters and memos.

What does it take to make an effective memo? Let me send you a memo on the subject.

May 27, 1981
To: People Reading THE WRITING SYSTEM
From: Edmond H. Weiss
Subject: *Better Memos*

The purpose of this memo is to give you the five main rules for writing effective business memos:

1. State the purpose at once.
2. Stick to one—or at the most two—subjects. Don't drift.
3. Use a lean, informal style.
4. Use headings, numbers, or "bullets," which will put some visual variety onto the page.
5. In the last sentence, tell the readers what you want them to do—even if it's nothing at all. If what you want them to do is the same as the purpose, then repeat it.

Post these rules above your desk and study them each time you write a memo.

There, that's the idea. Lean, concise, one page. Again, state your purpose immediately, and be sure you tell the readers to do something in the last sentence. Writing an interesting memo without providing the readers with a way to respond to your ideas is just cruel.

Seriously, readers need to know how to dispose of a memo, how to release the tension or to resolve the pressure it has created. If the memo has been about a serious or threatening problem, it may be important to tell the readers that all you want is to discuss it at the next meeting. (Otherwise the readers might think they are expected to solve the problem.) Even when all you want the readers to do is file the memo for future reference, say so.

And one more thing about memos: Write them so that the readers' eyes can keep moving southward. Do not force the readers to look up continually at the subject line ("as referenced above"). No one likes to read that way, so do not ask your readers to.

Letters are a more "polite" form of memos. Traditionally, memos are correspondence *within* the organization, while letters are to people *outside* the organization. As a result, letters have to be prepared more carefully, dressed more formally than your memos.

It is on just this point that letters get into the most trouble. Otherwise easygoing and plain writers turn into pompous royalty when they start their letters ("We have been advised of certain problems . . ."). Or lawyers ("Pursuant to the above-captioned reference . . ."). Or effusive apologists (We regret that, due to unforeseen circumstances, we have been unable to . . .).

You must never use any phrase or expression in a letter that is used exclusively in letters. (I list more of these forbidden terms in Chapter 10.) You must not put on that stiff, stodgy, unctuous manner that we associate with the typical business letter.

Rather, you should write in the same kind of language you use in your other writing. If anything, the style should be simpler, with shorter sentences and paragraphs. You should begin your letters directly with a statement of the subject or purpose or occasion of the letter. If you must refer to another communication, do so as gracefully as possible. Say

> Here are the data you asked about last Wednesday.

instead of

> Reference your inquiry of Wednesday last, below please find the data requested in your letter.

Use pronouns in your letters. Say *we, they,* and especially *you.* Say *I* if you are writing as an individual rather than as the spokesperson for an organization. (Most editors would prefer, though, that the first word not be *I.*)

Be brief, rarely exceeding one page. If you are writing to complain about something, keep your anger under control and tell your reader exactly what is wrong and exactly what it will take to satisfy you.

If you are writing to give bad news, do not take too long in "preparing" the reader before you get to the unpleasant part. In most business transactions, the most effective way to break bad news is to just plain say it, without a buildup and without a profusion of apologies and excuses.

(I have often found it easier to accept an unfavorable decision than to accept the lame and illogical reasons offered to justify that unfavorable decision. So, do not presume that every sensitive decision you discuss in a letter needs a justification or defense.)

When you are *answering* a complaining letter, try to answer promptly so that the writer will not have time to become more angry with you or your company. Be gracious and calm, even if the other person is abusive and vicious. Restrain your sarcasm; do not accuse, do not be excessively defensive or apologetic. Merely do what you can to rectify the complainer's problem.

In general, make your letters more memolike. Every year the typical American business letter looks more like a memo and less like a traditional letter. Today's letters tend to be typed in "block" form, with all the material lined up on the left-hand margin. Today's letters even have a "subject line," something that most readers like but that only recently has been considered acceptable.

And, increasingly, today's American business letter is doing without the "Dear" line and the "Sincerely yours" line. Some major corporations have dropped those affectations entirely.

As you can tell from these remarks, most of the problems with letters have to do with style, layout, language, typography. These are matters that, strictly speaking, should not concern you during the planning and organizing steps in the Writing System. I'll return to them later when the topic changes to editing and revising.

Before we reach that stage though, we still have a first draft to write.

THE DRAFT

8

Writing the "First" Draft

THE BLACK BOX

In the Writing System, doing the first draft is a "black box." That is, I can tell you what should happen just before you start the draft (I have been telling you for about seven chapters now), and I can tell you what to do with your first draft after it is finished (that's the next five chapters). But there is little to say about the drafting process itself.

Put simply, there are a million ways to write a first draft, and any of them will do if the result—the output of the box—is a chain of intelligible sentences, paragraphs, drawings, and miscellaneous notes that address all the topics in your plan and outline. As long as the result is adequate, and the time spent on it is reasonable, then any method is fine.

In other words, I do not really know how people write first drafts at all. I do not know how my own get written! If there is any magic in the world, it must play some part in the process that turns raw ideas and objectives (the plan) into comprehensible sentences and paragraphs. Of course, to restate the point I have raised several times already, *your first draft cannot be your last draft,* no matter how skillful a writer you are.

And before we go much further, I want you to be sure you agree with me about this. You will never, never, no matter how hard you work, be able to write five consecutive first-draft sentences without writing something awkward or unreadable. What is more, you will probably never be able to dictate more than *three* consecutive sentences without producing something clumsy or obscure.

If you think that somewhere there are good writers who can dash off a thousand words of readable memo without a backward glance, you are wrong. If you believe the professional writers who appear on TV shows and claim that they publish their first drafts without so much as rereading them, you have been taken!

First drafts can never be clean and clear. If, with practice, you are able to write first drafts with fewer problems than those you write now, then that means you will have more time for the finer points of editing later on.

But do not allow yourself to believe that your first draft (on any document longer than three sentences) can be your only draft. Do not even try for it, lest you end up spending more time on the draft than you should. (It will still need work, no matter how you struggle with it.)

Why is this true? Is not the final draft just a more refined rendering of the first draft? And, therefore, could not a more skillful writer produce a refined first draft?

No. Writing a first draft is vastly different from preparing a text for a reader. When you are composing (that is, writing the draft), you are struggling with your material and against yourself. You are grappling with ideas, solving puzzles, making decisions, and—perhaps most important—battling against the sheer physical stress of trying to write in the face of pressures, distractions, or even your own boredom and distaste. (Do not be ashamed. Almost everyone, including many professional writers, would rather be doing anything else than write!)

Far from scolding yourself for not being able to write a smooth, readable sentence the first time out, you should comfort yourself that it is in the natural order for first drafts to be clumsy and long-winded. For example, in a well-written, carefully edited sentence, the most interesting information will appear in the predicate of the sentence, which will almost always be *at the end.* Yet, in the normal manner of composing sentences, most writers will blurt out the main information in a sentence at the beginning! So, it is natural and normal for a first-draft sentence to read

> An electrochemical model is the most-sought-after explanation for mental illness. (first draft)

when, very probably, the sentence should read

> The most-sought-after explanation for mental illness is an electrochemical model. (final draft)

(Do not be confused about this example. Later, in Chapter 11, I shall explain about the correct order of sentences.)

Preparing your text for the reader, then, is a different process, using different skills and instincts. Editing and refining are not a struggle at all; they are a slow, deliberate craft where hurrying yields nothing but mistakes and mess.

In engineering terms, composing the draft is *quick and dirty,* but editing and refining the draft is *slow and careful.*

The two best things that can be said of the first draft are, first, it "covers" everything in the plan/outline, and, second, it is finished. And because I have already told you most of what I know about planning and outlining, my next task is to tell you how to get your draft finished in as little time as possible. The advice in the next pages, therefore, is less about outlines and language and more about the psychology of writing on the job.

And one more thing. From now on I use "first draft" and "draft" interchangeably. Strictly speaking, I want you to write only *one* draft. If you use the system the way I describe it, you will prepare *one* plan, compose *one* draft, and edit that *one* draft until it is acceptable. As I said earlier, if you follow this system you should never have to substantially revise or rewrite that draft; you should never have to throw away more than a page or two.

So do not tell me you have no time to do more than one draft. You are probably spending too much time on the one draft you are now writing—and could put that time to better use. (Or, if you are already writing that first draft at top speed, and *not* editing it anyway, you are probably sending junk to your readers and absolutely need to spend more time!)

TRAPS AND TIME-WASTERS

I will not retract what I said earlier. There are countless ways to write the first draft, and you can do it any way you like. Among these ways, however, are certain practices that slow down the draft and exhaust the time left for the later parts of the process. I am warning you about these traps and time-wasters because I know from years of experience and hundreds of clients how attractive they are and how destructive. Avoid these pitfalls:

- "Winging it"
- "Marination"
- Simultaneous revision
- Psychological stresses
- Environmental stresses

"Winging It"

No matter how much I harp and complain, there are still those of you out there who will write no plan or outline. Only a few of you will "wing it" on longer documents (more than three pages), but many of you will think that shorter pieces—the hundreds of memos, letters, and other one- or two-page messages that make up most of your writing—do not require a formal or written plan. "I've got the plan and outline in my head," you tell me. Sure you do.

Let us forget for the moment that I have actually seen memos in which the author changed his or her mind between the first and second pages. Let us forget about the scores of reports in which the author changed the names of system components between the first and third pages. Let us forget the numberless inconsistencies and contradictions that make authors look lazy and incompetent. Let us just worry about your *time*.

An unplanned, unoutlined first draft is destined to be a useless mess. In contrast, a "quick and dirty" first draft written to a thoughtful plan and outline is likely to be a *useful* mess. If you have written to a plan, a sound plan, then your draft merely needs to have its language cleaned up. If you have "winged it," however, there is a good chance that you have started wrong, finished wrong, organized your materials wrong, contradicted yourself, and failed to achieve your objective—assuming you even know what your objective was.

Of course, many writers claim that they cannot really decide what they want to say or how they want to organize their message without writing a draft. Many supervisors say that they cannot guide the writing of a report or proposal until they see a "rough draft." Doing an unplanned, improvised first draft *is* a legitimate way to plan. But it is also the most time-consuming and dangerous way.

If you "wing" your first draft, and in so doing decide what you really want to write, then the best that can be said for you is that all or most of it will need to be rewritten. (The stuff at the beginning is probably abominable.) More likely, though, your improvised first draft, with its three or four false starts, has used up all the time you have for the project and will be published or sent out as is.

"Winging" the first draft is a legitimate but luxurious way to compose. And because it uses so much time, it is almost always an inappropriate way to write on the job.

(Incidentally, over the years I have had to write reports for clients who couldn't decide what they wanted until after I had shown them a "rough" draft —most of which they invariably disliked and made me rewrite. I do not work for them anymore.)

If, moreover, you get accustomed to showing rough work to your superiors, they will over the years accumulate a recollection of countless poor memos and reports; their memories will manage to forget that these were just the "rough" drafts and not a true measure of your abilities.

"Marination"

Most writers say that they have trouble getting started. Of course, they mean getting started with the draft; usually they do not understand that writing starts with a plan, not a draft.

To some extent, having a well-designed plan and outline removes the greatest impediment to getting started: not knowing how to begin. In my experience, people who might stare at a blank letterhead all morning without writing a sentence will, if they scratch out a plan first, compose the letter in ten or fifteen minutes.

Even so, planning does not solve the problem entirely. We all have the problem of the blank page, the infamous "block."

The most dangerous solution to this problem is to wait for inspiration. I call it the "marination" approach because many of the people who espouse it (including some of my friends who teach writing) urge the writer just to wait until the ideas start to come, until the sentences start to form themselves in the mind, until the raw material in the outline has marinated long enough in the brine of the writer's creative juices.

This is the most "artistic" of approaches—favored by every professional writer who has no deadline to meet! (I have even heard some writers claim that when they compose their drafts they are merely transcribing whole pages they have already "written" inside.)

The "marination" approach to first drafts—waiting until your ideas are pickled enough to serve—is impractical for almost every engineer and scientist I know. There is no telling when the stuff will be ready, and, therefore, there is a good chance that the writer will be up against a fearsome deadline anyway. (Along the way the writer has wasted most of the time that might have been used for composing and editing.)

Again, "marination" is one of the thousand possible ways to put out a draft. But it is wasteful and risky. And it is not even the favorite approach of most professional writers. The majority of working writers (not just journalists but novelists and playwrights as well) write on a schedule and force themselves to cover some pages even when they feel uninspired.

Sometimes this forced writing is poor and immature (that is, undermarinated). But most of the time it is just as good as the stuff that bursts out after long thought.

In fact, most people who use "marination" as an excuse for not starting the draft are just stalling, putting off the inevitable day when they will have to go at it and get it done.

Simultaneous Revision

Composing and editing are such very different processes that it is inherently inefficient to try both at once. When I tell you that you cannot write a first draft that is good enough to be the last draft, I am also saying that you cannot revise and refine your first draft while you are writing it.

Yet many writers try to shift gears; they write a hurried paragraph and then take a deep breath and begin to edit and revise it. Several minutes later they struggle to pick up their train of thought and get on with the next paragraph. (And the paragraph they edited still needs work, by the way.)

Simultaneous revision cannot work. There are several reasons:

- The intellectual operations of composing and editing are so different that, in any given hour, you are unlikely to be in the right frame of mind for both.
- The process of switching from one mode to the other is distracting and stressful —especially when you notice how slowly the pages are piling up.

- The ability to edit well depends on being able to read the *whole* message; scores of questions about words and sentences cannot be answered until the entire draft is composed.
- The frequent starting and stopping so lengthens the time that the number of interruptions increases, making it harder for you to understand your own outline or remember what you wanted to accomplish in the paragraph you were about to write just before the phone rang.

As in the case of "marination," simultaneous revision may also be used as a stall or an excuse. As long as you are doing *something* to the report, you cannot be blamed for not having finished it. (I know some scientists and academics who have been pushed to publish findings before they really had anything to write; they like to play with the Methods section for several days in the hope that they will finally think of some way to present the uninteresting data they have generated.)

If you have always written this way (thinking that the result would be polished, final copy), it is a hard habit to break. The most irksome problem is when you know you have written something that is incomplete, unclear, misleading. You want to stop and fix it now! But you would do better to write a note in the margin reminding you to expand or improve this section later, when you are less hurried. What might be a five-minute repair job later on could cost you a half-hour of momentum on your draft now.

Try not to reread until you have finished the whole draft, or that whole day's portion of the draft. Better still, leave it alone until tomorrow.

Psychological Stresses

Writing a draft is hard enough without your making it harder.

Avoid certain psychological traps that drain your energy or hurt your concentration:

- **Do not set unattainable deadlines for yourself.** Do not, in your eagerness to please someone or finish with something, tell your boss or client that it can be done in an absurdly short time. Setting this silly limit will not speed up your writing; it may even slow you down.
- **When you realize you cannot meet a deadline, get an extension.** Most writers are more productive when they have a deadline, but no writer I know can function well when he or she is about to miss a deadline. Be realistic: Most of the due dates and milestones in business and government are *completely arbitrary,* and most delays have *no consequences.* As a matter of pride, of course, you want to keep your promises and meet your obligations. But there are times when you just cannot, and, until you get that extension, your nervous mood will prevent you from writing anything at all.
- **Do not be proud.** If you are stumped, say so. If you do not know how to write something—and cannot learn how in time—get help. If you are overcommitted, get support. If you are overloaded, delegate. If you are stuck, call for assistance.

The longer you delay admitting your problem, the sooner you will be up against an unrealizable deadline.

- **Do not run away.** Be honest about all those other things you "have to do." Do you really need to go see if the mail is in? Do you have to check your auto rental reservation for next week's meeting in Albuquerque? Isn't there some other time when you can have your hair cut? I once heard a writer say that every errand we invent when we are working on a draft takes a whole hour—even short phone calls and ten-minute conferences. This is especially true when your real motive is to escape the unpleasantness of the first draft. So be warned: I am on to you!

It is hard to know just what psychological state is appropriate for the "black box" of the first draft. Generally, it is good to feel the pressure to work, but bad to feel oppressed by the deadline. It is good to feel challenged and stimulated, but bad to feel overwhelmed by the scale or the complexity of the material. It is good to feel independent and assertive, but bad to feel that you are above the need for help.

Find your own balance. You are the only one who knows what will work best for you.

Environmental Stresses

When they teach technical writing in college, they neglect to tell you about the incessant din in which you will eventually be working. They overlook the telephones, the meetings, the folks dropping by your office to kibbitz, the boss who needs a lift to the airport.

They also neglect to tell you about fuel conservation programs that seal your office and make it like a sauna, or innovative architectural renovations that place you in a beautifully paneled room whose atmosphere is dangerously full of your own carbon dioxide.

On the positive side, they also ignore the effect of looking out the window on the first mild day in March or of smelling the coffee and fresh sweet rolls as they are brought down the corridor outside your lab.

In short, they do not prepare you for the hazards and temptations of your physical environment. You have to learn about them on your own. And you must learn to follow these rules:

- **Find a place that is as quiet and well lit and well ventilated as you can get;** do not try to write while breathing your own waste gases.
- **Head off as many phone calls and visitors as possible;** if necessary, find a time of day when nothing much is happening (like six in the morning if you can stand it).
- **Be sure you have all the supplies, paper, and equipment you need close by—before you start;** avoid the temptation of wandering down to the secretarial pool to borrow a stapler.
- **Find just that level of noise and activity best suited for your temperament;** learn how much quiet you can stand, and how long you can work all alone without getting nervous.

For most of us, the offices, labs, plants, and stores are not the best places to write. (Nor, for different reasons, are the forests and beaches.) No matter what, we will be disturbed by some factor that irritates us, interrupted by someone who craves our attention, or bemused by some person or image that lures us.

The less we like to write, and the more we dislike the particular report we are supposed to be doing now, the easier it will be to become distracted.

So protect yourself. If there are certain conditions that are intolerable for you, insist that they be changed. If there are certain temptations that overpower you, go out of your way to avoid them. And, above all else, write your first draft *quickly.* The longer you spend, the more you stall or worry over the syntax, the more likely you are to be interrupted or called away.

And the damned report will still be there when you come back—just waiting for you!

FIRST-DRAFT TECHNIQUE

The technique you use to finish the first draft does not matter—as long as you finish it. Study your plan; estimate the time you need to prepare the first draft; decide what parts you will do in what order . . . and go at it.

One of the variables, of course, is how much work you have already done on the plan and outline. If your outline is just a few key phrases, your draft will have more "new" material in it than if you had prepared a detailed storyboard with headlines, captions, and artwork.

For most of my drafts, there are only a few design questions. On short messages, one to three pages, I sketch a small outline and think very slowly and carefully about the first sentence. Once I know what the first sentence is (usually a crisp statement of the purpose of my message), the rest flows logically and naturally.

On longer projects I have to plan more carefully. My first concern is whether the draft will be driven by text or by exhibits. (By "exhibits" I mean all the figures, pictures, charts, tables—everything but ordinary sentences and paragraphs.) Although I have no firm data, I believe that more than half of my technical clients start with exhibits and write around them. This picture-oriented approach makes great sense in technical proposals and manuals. In fact, with the exception of short letters and memos, it is hard to imagine a scientific or technical document that could not be written around its illustrations.

So, even if you are not using the "storyboard" technique mentioned earlier, you may still find that the best way to bang out a draft is to sketch all the art and charts, arrange them in coherent sequence, and write enough sentences to tie them together. (This will not be good enough for the finished version, of course. But, remember, this is the *first pass.*)

If the nature of the project, or if your temperament as a writer, does not encourage this picture-oriented approach, then you may write sentences and paragraphs, stopping often to ask if an exhibit or a picture would help clarify the discussion. If you are unsure, or if you do not know what the best exhibit or chart should be, merely leave a note for yourself in the draft, a reminder that says, for example, "Put a flow chart here."

(The practical necessities of publication make it urgent that you develop your artwork as soon as possible. The sooner you are sure about the exhibits, and the sooner you get the necessary releases, the sooner this particularly slow part of the production process can begin.)

Sometimes a mixed strategy is the best approach. In many proposals, for example, there are sections that contain three of four exhibits for every thousand words of text; the work plan and schedule sections are a good example. In others, though, the ratio of pictures to text is smaller; the background discussion, for example, might have one or two charts in five thousand words. Thus, you may decide to use different drafting techniques for different parts of one large document.

(Remember, though, that most readers grow weary of uninterrupted text, especially if the paragraphs and sections are long. To the extent possible, THINK PICTURES.)

Once the text-versus-exhibits issue is resolved, the next technical question is *sequence of parts*. What order will you follow in writing the sections?

The safest approach, of course, is to start at the beginning and work through, but that is not always the best approach. For one thing, the beginnings of manuals and proposals are usually very different from the "guts" of the document because of their style and content. Saying good things about your software (the Introduction to the *User's Manual*) usually takes more writing skill than listing the control cards needed to generate a certain report. Many people who are good at the straightforward technical part of the book are stymied by the fuzzy, "flowery" parts that usually come at the beginning and end. If you are one of those writers, skip the Introduction and jump right into the parts that come easiest; do not use up your energy on ten elusive pages and leave no time or strength for the two hundred pages you can write effectively.

My own technique is to appraise each part of the document I am about to write. I estimate its length, and I predict the ease with which I shall be able to write it. My goal is to cover as many pages as possible, as easily as possible, because I know that one of the most debilitating psychological stresses is the feeling that time is running out and not enough is getting done.

Over the years, I have learned to save the easiest part for last, however. I want to have something that will flow out effortlessly when my stamina and concentration are nearly gone. Thus, I usually start my drafts with the "second easiest" part and work my way through the increasingly more difficult sections. When there is an apparent tie in difficulty, I do the longer of the two, with a view to covering as many pages as possible. When there is a tie between two sections of equal length, I follow the sequence in the document as my rule. I have found that I make fewer first-draft errors when I write in the proper sequence of sections, but this benefit is less important to me than the benefit of getting the greatest number of pages in the shortest time.

You may object to my reasoning. You may argue that the "hardest" sections might be the most important sections, and that by using up your strength and time on the easy parts you will leave insufficient resources for the toughest parts.

You may be right. What happens more often, though, is that you buy time to find an easier way to write those hard sections, or you find someone to help you with them—someone for whom they are easier. In other cases, you will get an extension on the project, using the large number of finished pages as proof that you are making progress and not just stalling. And, once in a while, you may even decide that those hard parts do not need to be written at all, that you were too ambitious in planning them, and that they seemed hard because you do not really know enough or have enough data to write them well.

Always be alert to any decision or procedure that stops the flow of your draft. Never claim "technique" as an excuse for not writing; never say, "I have to stop now because I haven't got the output I need to finish this section." Go to another section; keep moving.

When all else fails, leave blanks and "slugs." (Slugs are pieces of blank motion picture film or blank sound track used to hold a place in a film that is being edited. Eventually the slug will be replaced; for now, though, it lets you see the rest of the film.) Instead of agonizing over your inability to think up a good example or application, just leave a slug that says "ADD THREE SAMPLE APPLICATIONS HERE" and keep going.

I think you are getting the point now. The only good first draft is a finished first draft! Let nothing slow you down. If you are stuck where you are, go on to something else. (One of the main virtues of having a plan and outline, especially an outline made up of small "chunks," is just this: You *can* jump around and know that, eventually, everything you have written will find a place.)

Nothing short of genuine physical distress should stop the draft, and even a severe headache should be ignored when the cause of the headache is your feelings of guilt about not doing the report on time.

FIRST-DRAFT TECHNOLOGY

So much for technique. What about technology? There are four "ways" to generate a first draft:

- **Manuscript** (with a writing instrument in your hand)
- **Dictation** (which is then typed by someone else)
- **Typing** (on a conventional typewriter)
- **Word Processing** (that is, typing into an automated memory)

There is no aspect of your writing—and no aspect of this Writing System—more susceptible to change than first-draft technology. And there is no surer way to cheat yourself of savings in time than to ignore the machinery and mechanics with which you generate your draft.

Manuscript is the oldest and the safest way. People have their favorite pens and pencils and felt markers, and their favorite sorts of writing tablets. For as long as they can remember, a memo has been drafted on yellow, legal-size tablet paper, and always with a conventional-tip Flair pen—black only. (We become cranky and finicky about these preferences; I cannot write anything at all with a pencil, dammit.)

If manuscript is your choice for writing the first draft, observe these rules: *Do not write in pencil,* not because I do not like it but because it smears and disappears. Do not write with any instrument whose marks fade or rub off; do not work with any device that erases easily; do not write on any paper or surface that is easily distorted or defaced.

Write on lined tablets (not graph paper, unless you are making graphs and pictures), the wider the lines the better. Write on every other line, or even every third line; leave room for corrections and revisions. Write only on one side of the paper, and number the pages.

Use the largest handwriting you can summon up. Practice writing letters on extra-wide-lined school tablets, the kind that very young children use. If you can letter quickly ("print"), do so. If you only letter in the upper case (true of many engineers), then either make your capitals bigger or pray that your typist knows what capitalization is called for.

Dictation is the fastest technology for first drafts, or for the text part of the first drafts. Do not be intimidated; you are not supposed to dictate polished, perfect copy; you are putting out the first draft. (The only people who can dictate clean, readable letters and memos without editing them are those whose typists are skillful editors as well.)

Although most offices are equipped with about a half-dozen different recording devices (belts, wires, tapes of several sizes and speeds), the prevailing preference seems to be for the standard-size cassette, which can nowadays be used in recorders only slightly larger than the cassette itself. And because cassettes and recorders are everywhere, you can dictate anywhere, at any time. With some more-sensitive devices, you can even dictate in noisy rooms or airplanes and still be heard quite distinctly.

If you have not yet begun to dictate things, start soon. Practice reading your memos into the tape recorder. Work on making your pronunciation and articulation as clear and precise as possible. If you have a heavy accent, get help in correcting it.

When you have mastered the mechanics of reading clearly into a tape recorder, you should begin the process of reading less and improvising more. Write out the first and the last sentence of some short letters and improvise the middle. Or write just the first sentences in each paragraph of a report and improvise the rest. Over a few weeks, practice with less and less writing. Ultimately, improvise from your outline.

I hope this does not seem a heavy burden for you. Once you overcome your misdirected intention to dictate the final version, once you learn to speak more clearly, once you learn to dictate from an outline—once you have mastered these skills, you will be able to save thousands of writing hours over your career. You will mow down your routine reports and correspondence; you will convert "downtime" to productive time; you will be able to pick the places where you write.

Most of the advantages of dictation come from using dictation *equipment,* rather than stenographers. Even if you work in one of the few remaining places where secretaries can actually take shorthand, I still recommend machine dictation. Except for the rare stenographer who actually edits and improves your materials while transcribing and typing, stenographers are less accurate than recording devices. Moreover, recording machines are less intimidating to the writer (especially when you become confused and want to start over). And dictation machines will work any hours you want—or even come home with you if you ask them to.

(I happen to think, by the way, that most engineers should learn shorthand for themselves, as a way of taking notes at meetings or on trips.)

And one further note about dictation: Be sure to tell the typist that the tape or belt is *draft* and should be typed double-or triple-space, at *draft speed.* Recall that you intend to mark up this draft, and there will be no room for corrections if it is single-spaced. Also, this way the typist will bang it out much faster (saving *your* time) and be less chagrined when it comes back with a hundred changes in it.

(Soon, by 1990 or so, your dictation tapes may be typed by an automatic system. The manufacturers probably won't release them until they type *letter-perfect* renderings of the tapes. This is a silly constraint. Voice-activated typing can only be used for first drafts anyway, so why make us wait for refinements we do not need!)

Typing on a conventional typewriter is my own favorite way to make a first draft. Like many others, I was driven to typing by the illegibility of my handwriting, or, more specifically, by the urgent complaints of the typists forced to read my manuscripts.

Typing is easier than you think. Any one of you out there right now can type forty or fifty words a minute with an hour or two of practice. (Typing fifty words a minute *without mistakes* is difficult and takes months to learn. Fifty words a minute *with mistakes* is a snap.) There is no single letter or character that you can write or print with less effort than it takes to produce that letter on a typewriter. Typing has to be faster.

A typed first draft helps everyone. You, the writer, get a clearer view of how long the finished work will be. And your typist finds it easier to decipher just a few handwritten words (your corrections) than to struggle with the typical manuscript. And those typists who have some control over their time will choose the typed draft first—saving you still more time.

(I will excuse women professionals from typing their drafts, however. They have enough trouble as it is from people who presume that they are secretaries. They would do well to be seen nowhere near a typewriter, which instrument, as Margaret Mead once observed, first brought women into the business world and now confines them to a small part of it.)

One of the best reasons to learn to type is that more and more firms are asking their engineers and technical staff to type drafts directly into the *word processor,* a computer that holds the text in memory, plays it back for any corrections the authors might want to make, and then types a perfect rendition of whatever they have put on the memory device.

I have little doubt that very soon the *typical* way of producing a first draft (it is now rare) will be for the authors to type it directly into a system and make their own corrections on the system until it is acceptable. Right now, authors are expected to *type* the first version; in a few years, they will probably *dictate* the first version and make the corrections with the typewriter or electric pencil. In just the way that many mechanical designers are now generating their own final artwork with computerized graphics systems, most technical authors will soon be expected to generate their own typed texts using word-processing technology —which is becoming cheaper and more versatile every day.

Eventually word processors will help us all to write better; they may even include programs that flag our errors and improve our style.

That's the future, though. Right now, if you were to search the memories of the typical "intelligent" typewriter, you would probably find the same old stupid sentences.

ATTACKING THE DRAFT: SYSTEM STEPS

Figure 8-1 reminds you of the few, simple steps in the drafting part of the Writing System. Given a complete plan, approved by the necessary supervisors, the steps are

2.1 **Assemble Materials; Prepare Environment.** Make sure that you have not built in an excuse to go in search of missing data or supplies. Learn to situate yourself so that you are as free as possible from the distractions and environmental hazards described in this chapter.

2.2 **Write the "First" Draft.** Arrange the parts in a sequence that will ensure you the greatest number of pages in the shortest time. Attack the draft with energy and vigor. Forget style, forget elegance; look at your plan and outline and write as much material as you need to cover every point. Do not edit at the same time; do not wait for missing data; do not reread. Keep going until the draft is finished.

2.3 **Do the Interim Typing and Artwork.** Get your first draft typed at draft

2.1 Assemble Materials; Prepare Environment	• Get everything you need • Do not allow critical shortages • Block out as many distractions and interruptions as possible
2.2 Write the "First" Draft	• Arrange the parts in a workable sequence • Attack with vigor • Do not look back or edit
2.3 Do the Interim Typing and Art	• Type first draft at "draft speed" • Finish drawings as quickly as possible • Put approved artwork into production
2.4 Decide If the Plan was Followed	• Check to see if every item is included • Check to see if themes and strategies are included • Add first draft material, as needed
2.5 Set Manuscript Aside	• Get away from the manuscript • Put a physical barrier between you and the first draft

FIGURE 8-1 Attacking the Draft (Detailed Version)

speed (unless you have already typed it), with double or triple spacing. Finish as many of the drawings and figures as you can and start them into production —holding on to copies so that you can make sure your text matches your exhibits.

2.4 **Decide if the Plan Was Followed.** Check to see that the finished draft covers the material it was supposed to. See that no sections were left out (or badly underdeveloped). Make sure that the strategies and themes and general appeals you wanted to put into the message are there, at least in a crude but recognizable form. If not, add more first-draft material until you decide that the plan has been accomplished.

2.5 **Set the Manuscript Aside.** Almost every professional writer and editor agrees. You must allow an interval between the first draft and the revisions —as long an interval as your schedule permits. In fact, if you are in a first-draft mood (energetic, "quick and dirty"), you would do better to go on to the first draft of another message before launching immediately into the revisions of the one you just wrote.

At the very least, as I told you several chapters ago, try to put some *physical barrier* between the first draft and the revisions. Go for a walk, run, or swim. Change offices. Take a nap. Break the first-draft mood before you proceed to the editing. Otherwise you will not see what you need to see.

EDITING AND PRODUCTION

9

Editing I—The Process

STANDARDS FOR TECHNICAL AND PROFESSIONAL WRITING

Several times now I have compared writing to engineering, saying that the written product must be designed and executed in just the way that other complex products are created. I have also said that the better you design and execute your written products, the more likely they are to achieve your objectives and help you in your work.

The questions, then, are, How good is good? What tests or standards can I use when I write? Do all kinds of writing have the same standards? Is everything I write supposed to be perfect or exemplary?

Notice that these questions are identical with the questions raised in all aspects of engineering and science. Notice also that they create problems and conflicts for the writer, just as they do for the engineer or researcher. The writer asks, "It is all right for me to have a paragraph that contains thirty typed lines of text?" The engineer asks, "Must I double the joists under every partition?" And the answer is, usually, an unsatisfying "It depends."

In general, the standards for written products should increase, become stricter, as

- The number of intended readers increases
- The number of copies increases
- The distance the document will travel increases
- The "shelf life," or time of expected use, increases
- The familiarity of the reader with the writer decreases
- The sensitivity of the content increases
- The power or wealth of the readers increases

These factors never come up in high school or college composition classes. When we learn to write themes for our teachers, we are taught that everything we write must be as good as we can possibly make it—regardless of the time it takes, regardless of the relative importance of that particular writing assignment in the overall set of things we have to do. In that sense, what we learn about writing in school is impracticable for the job.

In the "systems approach" to editing, though, we never lose sight of the objective we are pursuing. We use the Writing System not to ensure that everything we write will be of extraordinary quality, suitable for publication in the most-demanding journals. No, we use the Writing System to ensure that no matter how much time we spend on the project, we can be certain that we did the best job possible within the constraints.

In some cases, the constraints are imposed on us: a fixed deadline, a strict budget. In other cases, we set our own constraints, deciding that we will not allow ourselves more than a certain time for a piece of work. (Usually this means that we have decided that some other task—possibly some other writing task—merits more attention.)

Once you begin to use the system, you learn how long it takes you to achieve various levels of quality. You learn how long it will take you to "clean up" a typical draft for

- Mechanical correctness
- Appropriateness of language
- Clarity
- Accessibility
- Urgency

Mechanical correctness is a concern for *rules.* You want to eliminate errors of spelling, punctuation, grammar, or sentence structure. (Usually, these are the last things you need to worry about in the final stages of editing and typing, along with general appearance.)

Mechanical correctness is the easiest standard to achieve. Most educated people know most of the rules, and the rules they do not know they can easily check in a reference book. As a result, there is little difference between the lowest acceptable quality for mechanical correctness and the highest.

No one should make spelling errors, although misspelling the word *affect* in a note to a friend will not do too much harm. All business letters should be completely free from mechanical errors; anything less is considered ill mannered. Proposals should be precise and correct, but readers of hastily written competitive bids will usually forgive those mistakes they consider to be minor typographical errors. (I once wrote a proposal in which the expression *Anchor Test* was garbled by the typist into *Granchun Test;* the proposal won, and no one ever asked me what the *Granchun Test* was.)

Put simply, almost everything you write should be free from all mechanical errors that are likely to be recognized as mechanical errors by your readers. Some errors—such as misusing the expression *due to* or misspelling the word *supersede* —are so common that few readers will notice. But do not take chances anyway; get everything right.

Appropriateness of language is a concern for the reader's skills and background, an attempt to use language that is familiar and comfortable for the reader. But this standard does *not* justify the use of dense, unreadable language when you write to a person who usually writes in that unfortunate way. My experience has been that even people who write "gobbledy-gook" do not like to read it.

Most important is the choice of words and terms that the reader will recognize and understand. A recurring problem for technical people is to find language that will be understandable to nontechnical readers, or to technical readers with a different background and experience from the writer's. (In an age of specialization, few people know exactly what *you* know.)

Clarity is the main and central standard. If your writing is mechanically correct, and if your language is appropriate, you have a chance to be clear. By clarity I mean resistance to misunderstanding; a clear sentence could not be misunderstood even if you hired a team of experts to derive some misreading.

Clarity must be defined negatively. If a sentence or passage is clear, then people will not misread it. If it is clear and asserts that A causes B, no reader will think it says B causes A. If it asserts an opinion that the reader agrees with, the reader will not think it asserts an opinion contrary to his or her own. If it instructs the reader to perform some operation, the reader will perform the operation as the writer intended.

The opposite of clarity is *ambiguity.* If a sentence is ambiguous, then there are legitimate ways to misread it. In the sentence

Every particle of matter in the universe attracts every other particle.

the phrase "every other" might be construed to mean *alternate,* as in "We bowl every other Thursday." The better rendering would be

> Every particle of matter in the universe attracts all other particles.

Clarity is the central virtue of professional and technical writing. If your writing is not clear, none of its other virtues counts for much. A well-written sentence is so clear that competent readers cannot misunderstand it. Let me illustrate with a masterly sentence from Peter Drucker:

> Profit and profit alone can provide the capital for tomorrow's jobs: not only for *more* jobs but for *better* jobs.

The best thing about this sentence is that it *cannot be misunderstood,* even by the people who disagree with it. In short, if you know the meanings of the words in that sentence (note that "capital" and "tomorrow" are the only three-syllable words), you *must* understand the author's claim.

Ambiguity can also result from misplaced or misaligned words and phrases. In the sentence

> Analysts who talk about the past *constantly* intimidate new programmers.

we do not know whether the analysts "talk about the past constantly" or, in contrast, "constantly intimidate."

In the sentence

> Write your evaluation of the plan on the attached form.

we are not sure whether we are to write on the attached form or, in contrast, whether the plan is on the attached form.

Often, the reader will eventually puzzle out the meaning of the unclear sentence. But if the burden is great, or if the final interpretation is more a guess than an interpretation, the sentence cannot be judged clear.

If your writing is unclear, it must be improved. And not only does this standard obviously apply to technical communication, I believe that it applies also to all business and professional writing—even to advertising and the more expressive messages. The best way to sell something is to write clearly about it. If your sales and promotional literature is vague, your customers will distrust your claims and doubt the quality of your products and services.

At the lowest acceptable level of clarity, your writing must be free, at least, from phrases and sentences that *invite* misreadings. At the highest level, your diction and arrangement of words must be fastidious, impossible to misread.

Because we rarely achieve this highest level of clarity in "the real world," the *accessibility* standard has become an important correlate to the clarity standard. Accessibility is the effort or difficulty needed to understand you, the quantity of rereading, searching for clues, interpreting ambiguities, even guessing, that you force on the reader. In general, five factors make your writing inaccessible:

- **Difficulty** — using big words and long sentences, thereby taxing the skills of the reader
- **Clumsiness** — putting words and phrases in the wrong places or wrong sequence
- **Wordiness** — using several words where one or two would do
- **Density** — jamming too much information into a single phrase or passage
- **Wilderness** — lack of markers, headings, paths, guides, artwork, typography, or other devices to help the readers find their way through your prose

Everyone agrees, engineers and editors alike, that good writing uses no more words than necessary to make its point, and that, in general, the shorter a passage is, the better—provided words and phrases needed for clarity have not been left out. Thus, no one could possibly prefer

> At that point in time we reached the decision to send him an invitation to our meeting.

to the sentence

> Then we decided to invite him to the meeting.

(Be careful, though. Never place conciseness above clarity. Anyone who would trim a few words from the Drucker sentence about profit would also reduce some of its impact and open it up to the possibility of misreading.)

Skillful writers trim their sentences and think carefully about each of the three- or more-syllable words. Do not make the mistake, however, of confusing simplicity with clarity. Short sentences are not necessarily clear sentences, and there are no awards for accessibility when your sentences, though short, are confusing or clumsy.

Accessibility is less a matter of following rules than of using "good style" —choosing and arranging words so that your message is no more difficult to read than it has to be, so that your writing seems lean and concise (but not clipped and abrupt).

Accessibility may be the most difficult standard to attain. Certainly it is the most time-consuming. Making your writing more accessible is a *craft:* You do it slowly and patiently, taking pride in what you are doing. But if you are one of those writers who just want to be finished, who cannot be bothered about the needs and convenience of the reader, you are not likely to make your writing accessible. You will let your passages—clear enough for *you* to understand—be understood only by those who are willing to struggle with your prose. You are in trouble!

At the very least, be aware of the importance of accessibility or "readability." If your writing is so complicated that only people who read at the "twelfth-grade level" can understand it, you have disenfranchised millions of readers. (Less than half the people who graduate from high school can read at the "twelfth-grade level.") By the simplest statistical deduction, it follows that the harder your writing is to read, the fewer people there are who will understand you without a struggle. And the less likely you are to achieve your purpose for writing.

For most writers, it is enough to pursue these standards of correctness, appropriateness, clarity, and accessibility. Those who can consistently write memos and reports that are clear and accessible, with appropriate language and with nearly no mechanical errors, will be among the most competent writers, and they need aspire no higher.

But for those who meet these standards and want to go higher, to be more effective, the standard is *urgency.*

Urgent writing is compelling, fluent, lively, fascinating. Its readers read with great attention, eager to know what comes next. Urgent writing is important and provocative, but not sensational or unprofessional. It is light and pleasing, but not frivolous. (The opposite of *urgent* is *dull.*)

At the very least, you can make your writing more urgent by arranging the material in the most psychologically effective sequence and by trimming away any material that slows the flow of your message. To achieve the highest levels of urgency, though, you need to work harder on the language (especially the verbs), sometimes spending several minutes on the choice of a word or phrase—an investment that most writers would make for only the most sensitive and important documents.

Urgent writing needs variety. A passage in which nearly all the sentences begin with the subject is not likely to be urgent. A passage in which all the sentences are simple sentences, or a passage in which the only connectives are *and* and *or,* is not likely to be urgent. Nor is a passage in which there are no personal pronouns or transitional phrases (like *for example*).

I say these passages are "not likely" to be urgent because I cannot predict for certain. Urgent writing takes not only technique (which I can teach you) but also talent (which I cannot). A talented writer could ignore the counsel in the paragraph above and still manage to write a fascinating report.

For most engineers and scientists, though, the pursuit of urgency does not seem to be worth the candle. It takes too much time and does not give enough marginal advantages when it is achieved. In most cases, they decide that clear, accessible writing is good enough to achieve their objectives.

I encourage you to aim for urgent writing, as your final goal. But be sure that you can crawl (correct writing) before you walk (clear writing), and before you run (urgent writing).

3.1 Edit for Conformity to Strategy	• Review opening • Check organization • Test against outline • Check for clarity of purpose and suitability of strategy • Expand, alter sections--as needed
3.2 Do the 80 Percent Style Revision	• Work either top-down (logical) **or** bottom-up (practicable) • Edit until "diminishing returns" sets in
3.3 Show Version 2 to a Colleague or Editor	• Give 80% version to an "intelligent, naive reader" • Decide what still needs to be done
3.4 Make Final Style Changes	• Make final changes or • Entrust final changes to reviewer/editor
3.5 Secure Final Approval	• Show "final" version to superior • Change as needed to get approval
3.6 Text Production	• Make MS suitable for typing • Type
3.7 Art Production	• Communicate needs to graphics people • Approve draft versions • Produce final art
3.8 Final Assembly; Copy Editing	• Pull the materials together • Review for "bugs" • Double-check corrections • Prepare for printing or reproduction

FUGIRE 9-1 Editing and Revising (Detailed Version)

STEPS IN THE "EDITING AND REVISING" PROCESS

Figure 9-1 is a more-detailed version of the figure you first saw in Chapter 2. In it are the eight main steps you should follow in the Edit and Revise part of the system.

Step 3.1: Edit for Conformity to Strategy

Read your first draft again. (Remember, you should have put it aside for a while.) But do not yet read it for small matters of style and language. Read it for strategy.

Does it begin well? Could the reader say So what? or Who cares? after the first sentence or two? Is the purpose of communication expressed succinctly at the opening? If not, is there a good reason for *not* expressing it?

Look to the organization. In addition to starting at the right place, does it then proceed in a logical sequence? A sequence that is appropriate for the audience? Does the logic show? Is the skeleton on the outside?

Have you followed your own outline? If not, why not?

Have you taken my advice on planning in general? Have you followed my pointers for reports, proposals, papers, and so forth? If not, did you have some reason for taking a different approach? Are you confident about your decision?

Are your informative messages especially accessible? Have you done anything to control or reduce the effects of noise and distraction?

Are your persuasive messages argued cogently? Do you build a case and make good use of the various forms of "proof"? Do you adapt to your audience and make it easy for your readers to agree with you?

Are your motivational messages likely to overcome the readers' inertia? Do you offer inducements to your friends? Do you know if these inducements are the most appropriate or attractive? Have you used sanctions to threaten your enemies? Do you know if these sanctions are the most appropriate or intimidating?

In general, have you distinguished among informing, persuading, and motivating?

Have you included in your message only those materials that are relevant to your purpose or necessary in helping you to overcome any barriers between you and your readers?

Notice that all these questions are the strategic questions you should have asked at the very beginning. Now, in Step 3.1, you should be double-checking, making sure you did what you set out to do. At the very worst, you should only have to write a bit more—perhaps a new opening—or rearrange a few sections to ensure that your first draft is *strategically* correct. Perhaps you might want to expand a section that is not sufficiently developed for your reader, or shorten a section where you got carried away with writing about your favorite subject.

Step 3.1 ends when you decide that you have written *all of the right message.* The only defect in your draft is that it is rough, dirty, hard to read. A "useful mess." Now you take care of that.

Step 3.2: Do the 80 Percent Style Revision

Step 3.2 is the hardest part. You must examine your document in minute detail and make every improvement in words, phrases, and sentence and paragraph structure that you can.

Before an editor sees what you have written. Certainly before your boss or audience sees it. Before any "peer" reviewer sees it. Before anyone but you and the typist get to see it, clean it up. Make 80 percent of the improvements that need to be made.

Now, I know you are thinking, How can I tell that I have made 80 percent of the improvements when I have no idea what 100 percent of the improvements should be? The answer is, When you have made all the improvements you can find without help, when it has been several minutes since you were able to correct or refine anything, you are probably near the 80 percent point.

Editing, like most difficult jobs, works on the 80-20 principle. You can do

80 percent of the work in the first 20 percent of the time. After that, diminishing returns set in, and it becomes increasingly harder for you to find anything more to fix.

Some writers, of course, cannot get close to the 80 percent level. At least not without the kind of advice this book provides. But even the best—even professional writers—cannot see all the improvements that are needed without spending four or five times longer than they spent on finding the easy, obvious ones.

(Some writers circulate their unedited first drafts for comments and wonder why so many people do not understand the document. Others turn their rough drafts over to an editor, leaving it up to this person to clean up the mess, and they wonder why the editor "changes the meaning" of what they have written.)

There are dozens of ways to do the actual task of editing. Logically, you should start with the biggest questions first: Are the paragraphs (or sections) complete and coherent? But most writers do not edit logically; they start with the easiest part, some task like substituting good words for bad words. I call this approach illogical only because you will occasionally spend time reworking a sentence that, had you started with a bigger question, would have been thrown out completely.

Yielding to practical necessity, I have organized the next two chapters in the sequence that most writers follow: words-phrases-sentences-paragraphs. The advantage of starting with the easiest improvements is that, even if you do not have enough time to do everything you need to do, you will have made the greatest number of improvements. And these easy improvements are often quite significant; changing *initiate* to *start* and *facilitate* to *help* can be a dramatic improvement.

Step 3.3: Show Version 2 to a Colleague or Editor

As I said a long time ago, one of the worst time-wasters in writing is trying to do it all alone. At this point in your revising you have an important choice to make: to go it alone or get help.

If the piece you are working on is not very important and does not merit extensive editing, then decide to call it finished and send it in for final typing. If it is an important message, though, and you decide to finish it by yourself, expect to take considerably more time on the last few improvements than on the first several. And resign yourself to the fact that, no matter how hard you try, *you will be bound to miss things that another reader will spot immediately.*

I recommend that you show what you have written to someone else for two main reasons:

- First, another person can more quickly see the problems that are hard for you to spot. Each of us has certain linguistic peculiarities, bad habits, and quirks.

Your reader's will be different from yours and, therefore, your reader will spot yours right away.

- Second, the writer of a piece cannot ultimately answer the most fundamental editorial question, Is this clear? The writer understands the draft, sees things that are not there, fills in links that are missing, cuts through irrelevancies to the main point. The writer supplements what is on the page with a vast data base of information that is not on the page. All of us inevitably write passages that are COIK: Clear Only If Known. Only another reader who does not already know what we are trying to communicate can tell us when we are clear.

So, every writer who works alone will send out either material that is not nearly finished or material that is nearly finished, but not quite. (The latter will take much longer than the former.) A writer who works alone, without an opinion from another person, will inevitably and inescapably write sentences and passages that are unclear. Sentences and passages that could have been made clear if another reader had asked the right questions.

I urge you to find a colleague (if you cannot get an editor) to review your 80 percent version (Version 2). Choose the person carefully. Find an intelligent reader who knows as much as the intended audience but is naive (that is, did not work right alongside you on the project you are describing). A colleague, a friend, a spouse, a secretary—find any intelligent, naive reader you can. Reciprocate by agreeing to be the intelligent, naive reader for the other person.

DO NOT SHOW THIS VERSION TO YOUR BOSS, TO YOUR SPONSOR, TO A "PEER REVIEWER," OR TO ANY OTHER PERSON WHO IS IN A POSITION TO JUDGE YOUR WORK OR TO SHAPE YOUR CAREER. Just as I have a hard time convincing people that they should not write alone, I have an even harder time convincing them *not* to show unfinished work to critical readers. Despite the pressure on you to release reports and documents before they are ready, despite the sometimes considerable pressure, resist. Even though your reader (boss, customer, project officer, client) tells you that he or she will be forgiving and understanding, don't believe it.

In my experience, no reader—except a very good editor—can remember throughout the reading of a document that it is unfinished and that the author intends to improve it. Quite the contrary: Every unclear sentence, every weird claim, every abrasive phrase, every factual error, even every minor error of grammar and punctuation—every one is etched in the memory of that reader. Over the years, the effect of showing unfinished work to the same client or supervisor is to create the impression that you are, at best, a B− performer.

Even when your R&D contract calls for a "draft final report," be sure that what you submit is, in your judgment, A+ quality. If you relax, knowing that the client or sponsor is going to make a lot of changes anyway, you will have seriously damaged your future relationship with that reader.

So, on the one hand, I am insisting that you show your "Version 2" to another reader, but I am cautioning you against showing it to anyone whose opinion of you might affect your income or career progress. Heed both these warnings.

Step 3.4: Make Final Style Changes

Usually, you will take the questions and advice of your second reader and incorporate the needed final changes into Version 2. In some cases, though, you may elect to have your reader make the improvements for you.

This may strike you as a bold and risky policy, but I have found it to be the most efficient way. So long as you keep the rights to the final say, and are guaranteed a chance to see what the other person did, you really have nothing to worry about.

(In this connection, be warned that most magazines reserve the right to edit and revise your articles without your permission or review. Only the most serious periodicals—that is, the ones least worried about meeting their publication deadlines—guarantee the author the right to review the "galleys" of the paper.)

Step 3.5: Secure Final Approval

If you are the final authority on the acceptability of your message, that is, if it is your *own* memo or report that you are writing, then you are nearly finished. You may put your draft (Version 2, plus some extra refinements) into production.

Often, though, you will need one more approval—either from your immediate supervisor or from some other person whose name will appear on the document, or sometimes from an officer of the firm. (For proposals, there may still be the red team to contend with.)

So, you show your presumably finished version to the appropriate reviewer. And I hope that this reviewer is the same one who reviewed and approved your plan many steps ago. I especially hope that the reviewer is not getting his or her first glimpse of your work, that is, a first notion as to what you intended to do and how you proposed to do it.

In sum, there should be no surprises for the reviewer. At worst, the reviewer might challenge a phrase or two, perhaps suggest that certain passages be made stronger or clearer (or weaker and vaguer, more likely). If the reviewer tells you that something is significantly wrong—that there are "serious problems," as government people usually put it—then you must have done something terrible along the way.

If you have followed the system, it is nearly inconceivable that you could have written a piece that is substantially wrong, something that needs—gulp!—a Version 3.

To be in this dilemma you must have neglected some important step, probably planning, and probably with the excuse that you lacked the time to plan. Now see how long it takes you to compensate for a lack of planning!

But I am describing the worst case. Typically, if you have used the system, it will take very few adjustments at this step to make the version acceptable.

If the document is long and will take some time to review, try to get an earlier approval of the artwork and tables. These are the slowest parts of the writing and production process, so you should try to get the graphics under way while you are still refining the text. (Be sure to keep copies, though, to make sure that your text is consistent with these graphics and exhibits.)

Step 3.6: Text Production

Now it is time for final typing. And, as I said before, the way you submit your manuscript to the typist may affect how long *you* have to spend finishing it.

If Version 2 is in longhand, be sure it is legible. If necessary, rewrite some passages. If there are many inserts ("see insert B"), try to "cut and paste" the draft so that the inserts are in the right place.

If there are any special typing instructions—rules for headings, formats, unusual conventions—tell the typist in advance. Make sure the typist knows precisely what you want, especially if what you want is unusual.

In some cases, of course, the final typing will not be the work of a typist at all, but of a word processor. Possibly you, the writer, will give the final instructions and put them into the typing system. This may be the best approach; it ensures that you will get what you want, and it gives you one last chance to make small improvements. And even though many of my manager clients claim that they cannot "waste their time" typing, I suspect that many people really have nothing more worthwhile to do than to make sure that their documents are just right.

Do nothing to irritate, frustrate, or abuse the typist. Illegible, "untypable" manuscripts waste *your* time and lower the quality of the finished product.

If your message is to be typed "camera-ready," that is, suitable for photo-offset printing, be sure you or your typist knows the correct typing "specs."

Step 3.7: Art Production

Artwork—pictures, graphs, tables, charts—is the heart of much technical writing. A typical electronics proposal will have as many pages of diagrams as of text. The final report of a testing project will have more exhibits and tables than paragraphs.

The keys to effective artwork are

- First, be sure your firm has the talent and the equipment to produce high-quality art and graphics, either "in-house" or from a contractor.
- Second, be sure that the people who have to do the exhibits and graphics have enough time to do a respectable job; *artwork cannot be rushed.*
- Third, tell the people doing the work exactly what you want, using the proper terminology. If an artist uses a term you do not know, ask for an explanation.
- Fourth, when you need ideas, get advice from your graphics people. Often, a

typist will tell you the cleverest way to set up a complicated table or lay out a questionnaire. The artist may suggest a lively way to depict your system, or the printer will show you a way to make a cluttered page more accessible. Take this advice—but be sure you know what you have agreed to.

Step 3.8: Final Assembly; Copy Editing

Now it all comes together, literally the culmination of all your work. Are you finished? Not quite.

Read your message once more. Read it closely, meticulously. Even though you are weary of it by now, even though you may be falling behind schedule, read!

Look for *bugs.* Search for the small errors of grammar; rethink your punctuation; verify the spelling of terms that look strange.

Whether these bugs were in your original version or whether they were brought in by the typist does not matter. *You* are responsible for them. *You* will pay the consequences if they leave your desk uncorrected.

Of course, some writers delegate this part of the job to a secretary or assistant. That is fine—if you have confidence in the language skills of the person. (Some firms do hire secretaries in part because of their editorial skills; most do not.) At least be confident that the secretary or assistant knows as much about your language as you do.

In most cases, the time you spend on checking the typist's version is time well spent, a few minutes that will save you from making an ass of yourself. And an equally important task (for the same reason) is *double-checking* the final version to see if all the bugs and errors you spotted earlier were, in fact, corrected.

I am always amused at the faith we have in typists, a nearly religious certainty that typists—unlike every other class of employees—follow instructions perfectly. Only this profound faith could account for the practice of marking up a text with final corrections and never double-checking to see if the corrections were made at all, and made correctly.

Typists, you see, are just like the rest of us; they miss things, make mistakes, and occasionally cut corners and goof off. Just because you circled a misspelled word does not mean that the typist noticed the circle, or that the typist knows the correct spelling, or that the correct spelling will have been typed in. Indeed, there is no way of knowing in advance whether the typist, in the process of correcting errors, has not introduced some new errors!

(Word processors, of course, are too unimaginative to invent new errors while they are correcting old ones. But then, word processors, unlike human typists, cannot work when they are sick.)

As boring as it may be, as much as you would like to be on some other project, stay with your written product to the very end. If you are a poor proofreader (as I am), if you cannot proofread the numbers in a large table (as I cannot), then get some help.

And if your document is to be *printed* (rather than photo-offset), find an editor who knows how to "type-spec" your manuscript before it goes to the printer. Be clear about every detail of type style, size, and layout.

Do not let your good work be ruined by a small, stupid oversight.

TESTING FOR READABILITY AND QUALITY

At any point in the editing and revising process, you can stop and assess how far you have come. You can test the "quality" of your first, rough draft, or any later version. Most writers assess their drafts intuitively; they "sense" that a piece needs work, or that it is too difficult, or that it does not "flow" well.

If you want, you can also test your draft *formally.* You can take the whole piece (if it is short) or a sample of two hundred or three hundred words (if it is long) and run statistical tests of its accessibility and "quality."

The most widely used measures of accessibility have been with us since the late forties and early fifties: Robert Gunning's *Fog Index* and Rudolf Flesch's *Readability Formula.* Both measures assert that two factors—the length of the sentences and the length of the words—are the most predictive of how difficult your writing is.

In the Fog Index:

.4 [(avg. words/sentence) + (percentage of hard words)] equals the "grade level" of difficulty required to read the passage with understanding. ("Hard words" are words with three or more syllables, with certain exceptions. See Robert Gunning's *The Technique of Clear Writing.*)

In the recalculated Flesch Formula (recently revised by the U.S. military), grade level is measured by

.39 (avg. words/sentence) + 11.8 (avg. syllables/word) − 15.59

Table 9-1 shows the Fog Index and the Readability Formula applied to the same passage of text.

Both formulas, the Gunning and the Flesch, have survived for several decades because they are remarkably useful. They are more accurate and reliable than you would expect—especially considering how simple they are—and they give a measure that makes sense to most people, even though most of us have only the vaguest notion of what it means to read at the "ninth-grade level."

Both indexes have been used to prove conclusively that mature, interesting, technically sophisticated writing in the better journals need not be difficult to read. And also that most of the dull, pedestrian memos and letters we read every day are at least 30 or 40 percent more difficult than they need to be.

But both indexes measure one thing only: *difficulty.* And that difficulty is narrowly limited to the issues of long sentences and big words. Neither index assesses clarity or interest or variety. Neither measures the quality of the writer's thinking, the craft of the writer's syntax, or the correctness of the writer's grammar. The Flesch Formula and the Fog Index measure difficulty alone, and diffi-

TABLE 9-1 Two Readability Formulas

Specimen Passage from a Manual ("hard words" underscored)

In order to best serve the total <u>customer</u> base, this <u>manual</u> is written in two levels. The first is a very basic <u>introduction</u> to the Techno <u>family</u>, and the second level is for the user who has to refer to the <u>manual</u> on more than an <u>occasional</u> basis and who wants to <u>rapidly</u> scan and find <u>specific</u> sections. For the user who is quite <u>familiar</u> with <u>programming</u> and the Techno <u>instruction</u> set, the <u>appendices</u> are the best <u>reference</u> in the sense that all the data which is discussed in detail in the <u>manual</u> is <u>summarized</u> in a series of tables for <u>convenience</u>.

Fog Index

Total Words = 102
Total Sentences = 3
Total "Hard Words" = 16

Average Words/Sentence = 34 (102/3)
Percentage of Hard Words = 16 (16/102)

Fog Index = .4(34 + 16) = 20th grade level

Recalculated Flesch Score

Total Words = 102
Total Sentences = 3
Total Syllables = 155

Average Words/Sentence = 34 (102/3)
Average Syllables/Word = 1.5 (155/102)

Flesch Formula = [.39 (34) + 11.8(1.5)] − 15.59 = 15th grade level

culty of only two kinds. Moreover, after you read the next two chapters, you will see how *easy* it is to drop the Flesch or Fog scores by two or three grade levels without much work at all!

Unfortunately, there may occasionally be a conflict between readability (how easy it is to read) and urgency (how interesting it is to read). Simple sentences (one main clause) are easier to read than complex sentences (one main clause, one or more dependent clauses). But writers who stick to simple sentences (and their close cousins the compound sentences, two or more simple sentences attached) are severely limited in how clever, persuasive, and urgent they can be.

My point is that if you take a systems approach to editing, you will be ever mindful of your purpose and audience. If your objective is to communicate facts to a machine operator who does not read very well, you will want to write in a choppy, simple style that scores no more than 6 or 7 on the Fog Index.

If you want to write something more provocative, however, and if you know how to write a longer sentence that is grammatically well formed, then the Fog Index may be less relevant.

Most of the people who teach writing to engineers and scientists do not expect much from their students. They recommend the use of short, simple sentences, starting with the subject, avoiding any complicated syntax. They know that the writing of students who follow this advice will be less dense and difficult and that it will contain fewer clumsy mistakes. They want their students to be clear, competent, adequate, understandable writers. This is certainly a desirable objective.

Beyond that, though, there is much more to be learned. And, two or three chapters from now, you may decide that you want to be more than adequate.

PREPARING TO EDIT

To edit you must be calm. You need not shut out all distractions and interruptions—as you did when you were composing the draft—but you do need to control the pressures.

You cannot rush editing and proofreading. You cannot skim; you cannot give it a "quick and dirty" once-over. If you feel nervous and impatient, do something else till you feel calm again. (Did you take that break after finishing the first draft? If not, you are in no shape to edit.)

Editors need very little to work with. (You are now the editor, remember.) You will want a red pencil, of course; scissors and tape; and perhaps a pot of the glue that typists use to "cut and paste." If you are going to be marking the "original," then you will want one of those blue pencils that are invisible to photocopiers.

Clear your desk of the debris that gathered while you were writing the first draft. Have a substantial dictionary near by, and perhaps a style guide.

If you are editing your own work, change anything that you think will help you to achieve your communication objectives. Add whatever you can to help the readers know, believe, or do what you want them to do. Remove or rework anything that will hurt the clarity, correctness, and general readability of the piece.

Picture your reader while you work. When in doubt about a change, imagine what your reader would think of it. Refrain from writing anything that would look odd or incorrect to your reader—even if you happen to be in the right. Think of your purpose, think of your reader, and let these two ideas be your guide.

And one more thing. When you are editing someone else's work, be kind. Nowadays, senior engineers and scientists have the privilege not only of writing and editing their own work but also of editing the work of their junior colleagues.

When you edit the other person's work, though, go easy. As long as it will be *his* or *her* name on the piece, not yours, you are obliged to change as little as you have to. You may correct errors, untangle the more-twisted sentences, replace inappropriate words with the appropriate ones, and so forth. But you must stop short of trying to make it more interesting and lively—unless that is the clear understanding between you and the writer.

If you are in the habit of substantially revising the work of your juniors—without showing them what is wrong or teaching them how to do it better—you will be encouraging them to depend on your editing skills. Not just for the last 20 percent of the improvements but for 100 percent. You will encourage them to do no editing and revising of their own. And you will leave them mystified when their employee appraisal form describes them as bad writers, that is, "not sufficiently skilled in written communication techniques."

Be gentle with the drafts of others. But when it comes to your own, show no mercy!

10 EDITING II— Words and Phrases

USING THE BEST WORDS

Fortunately, the easiest editorial improvements produce the most noticeable benefit. Finding the best words consists mainly in applying a few simple principles and making a few simple replacements. You don't even have to think very much. Just follow this advice:

- Do not show off.
- Write precisely what you mean.
- Do not waste words.

Secure people *do not* have to *show off* when they write. They will not choose a long, fancy, unfamiliar word when a short, plain, familiar word will do. This is *not* to say that good writers write in little words; they do not. Rather, they will look for the shortest, plainest word that means what they want to communicate.

This point is the simplest and most fundamental. Too many engineers take my courses because they want to write in a more "flowery" style—or, more precisely, because they think their bosses want them to write that way. But no one ever uses the word "flowery" to describe writing he or she likes to read, in technical communication or elsewhere.

Showing off, using bigger words than you need for the effect, is a transparent device. Generally, the less content or substance in your writing, the more you try to show off.

But it does *not* work! Little ideas decorated with big words look the way little children do when they dress up in their parents' clothes—ludicrous.

Good writers also value *precision* in their choice of words. They dislike ambiguity, and they particularly dislike the use of big words in contexts that make no sense. For example, a good writer would never call someone's manual "atrocious," or call a guess a "determination," or speak of a "slight commitment," or call something "relatively unique." These uses may be understandable, but they are imprecise and distracting to a reader who knows the language.

Good writers *do not waste words.* Nearly all first drafts are long-winded and wordy. As I have said before, wordy constructions slow down the flow of information, giving you more time to think while you write. But good writers look for certain familiar, wordy expressions and replace them with single-word substitutes. Improving is easy; each of us has his or her own list of offenders, two or three dozen phrases and expressions that crop up repeatedly in first drafts. Once we know the offenders, removing them is simple.

This part of editing involves no theory and almost no hard thinking. Once you agree that showing-off is childish, once you learn to stop—for just a second—and think about what your words mean, and once you learn to spot your particular list of wordy and windy phrases . . . you will be able to make dozens of improvements in your first draft.

Showing Off

Most technically trained people have odd notions about simplicity. Some think that being simple means being simplistic, or overly simple. (Be careful; many of my clients confuse *simple* with *simplistic.*) Some think that simplicity bars them from using technical terms; others think it forces them to write in a series of short, primer sentences.

The worst misconception is that simplicity resembles stupidity. One of my clients, for example, once told me that if he were to write the way I recommended, "No one would know I went to college!"

Most of the young engineers and researchers I meet insist that their bosses would reject (or edit mercilessly) simple reports or proposals, that writing them would work against their career interests. Some even say that their bosses hope that my seminars will make their writing fancier!

(Interestingly, every "boss" I have worked with tells me the opposite, that he wishes that people would write more clear, concise messages.)

Most of these impressions are naive and self-defeating. Solid ideas, clearly and simply presented, *never* seem simplistic or naive. Only weak, incomplete, careless work seems to benefit from overblown language—and not much at that.

Putting it bluntly, showing off by using unnecessarily large words is the tactic of lazy or incompetent writers. It distances their readers, bores their associates, ensures that their materials will not be read carefully. It is the tactic of people who do not want to communicate!

Consider these common examples:

FANCY	PLAIN
Nouns	
implementation	start, use
commencement	beginning
finalization	end
interaction	talks
utilization	use
indication	sign
requirement	need
application	task, job
condition, situation	state, status
compensation, remuneration	pay
capability	ability
reservation	doubt
conceptualization	idea, draft, plan
Verbs	
furnish, provide	give
utilize, employ	use
formulate, fabricate, construct	make
inspect, ascertain, investigate	check
possess, maintain	have
offer, present	give
indicate, reveal, present, suggest	show, tell
transmit, communicate, disseminate	send
require, necessitate, mandate	need
initiate, commence	begin
finalize, terminate	end
represents, seems	is
modify, alter, redesign	change
realize, appreciate	know
endeavor, essay, attempt	try
select, opt for	pick
retain, preserve	keep, save
acknowledge	agree, grant
establish	show
effectuate	cause
evaluate, appraise	test, rate

Do not misunderstand. The words on the right should not always and automatically substitute for the words on the left. There may be cases where you mean *preserve,* not *save* (a distinction we shall explore below). But most of the time, eight or nine times out of ten, the word on the left is merely a showoff's way of saying the word on the right.

This list, of course, contains only a few samples, some of the more-common offenders. There are hundreds of instances, in every part of speech. There is, for example, no legitimate reason to write *advantageous* rather than *helpful,* or *subsequently* rather than *later,* or *belatedly* rather than *late.*

The excuses that people have for using the longer forms are ingenious and amusing. The short terms are too blunt, I hear. They sound too conversational and informal. They are not professional or businesslike. (To do what I am proposing, they say, would get them into trouble.)

Bunk! We are so used to this pointless showing off that we believe it should be the standard tone for professional communications. In fact, *almost no one notices what you have done if you begin these improvements* . . . unless your phony language was hiding some mistakes or defect in your thinking. In that case, you deserve to be exposed.

Utilize and *utilization* are possibly the worst cases. I have four dictionaries, and each gives me a different ruling on the meaning of these terms. I have concluded, therefore, that the word *utilize* has no consistent meaning of its own and that we should all abandon the words *utilize* and *utilization.*

Notice that even though not all fancy words have three or more syllables, most words with three or more syllables are likely to be fancy words. For that reason, when I edit my own first drafts, *I presume against any words with three or more syllables.* That is, before I will leave any such word in my text, I try to talk myself out of it, convince myself that there is some simpler word that means precisely what I want to say. If the suspect word passes this screening (which takes only a moment, by the way), I let it stay. Otherwise I replace it.

To make yourself more efficient, keep a record of the fancy words that appear most often in your drafts. Compile your twenty favorites and put them on a chart above your desk. (If you have a word processor, you can instruct the system to flag or underline these terms, or even to replace them automatically.) For most of us, just fifteen to twenty-five fancy words account for 80 percent of our offenses.

Of course, my clients will sometimes tell me that they want to use these longer words for "variety," to avoid repetition. Yes, in some cases you may want to substitute *compensation* when you have written *pay* too often.

But be alert to two issues. First, repetition of key terms in a technical communication is not a sin. It is far better to be "redundant" and precise than to introduce a synonym that is not really a synonym. (The thesaurus, incidentally, is a dangerous collection of misinformation about meanings.) Second, most of the annoying repetition in technical and business writing is due to the *faulty sentence structure* and should be corrected with better sentences rather than with phony synonyms. (More about this in the next chapter.)

There are, of course, still other ways to show off. Some writers like to flaunt their meager knowledge of foreign languages. Unless there is a compelling reason to do otherwise, however, write in English. And if you use foreign expressions, use them correctly. (If you speak them, pronounce them correctly. The *fait* in *fait accompli* rhymes with bet, not bait.)

Avoid most foreign terms and abbreviations, especially

i.e.	re
e.g.	a priori
etc.	fait accompli
per	cause célèbre
qua	raison d'être
via	

If you must use *e.g.* and *i.e.,* remember that *e.g.* goes before a list of examples, not *i.e.*

Still others like to make ordinary ideas sound technical and complicated by the invention of new "jargon," the most hideous example from the 1970s being the word *prioritize* for *rank.*

Jargon used to be the name we gave to all gibberish or unintelligible writing (what we now call "gobbledygook.") Today, though, jargon usually refers to the overuse of technical and professional vocabulary in places where it does not belong. Or, worse, to the habitual creation of words and phrases that sound technical but are, in fact, nonsense.

Most of the jargon to be edited from your first drafts is the result of big words, tired phrases, legalese, too many nouns, and a few other assorted flaws that I shall tell you about later. Some jargon, however, is the result of pseudoscience, the invention of fake technical words like *definitize* when you mean *specify* or *refine;* or *descope* when you mean *shrink, limit,* or *contain.*

This kind of language—usually created by adding prefixes or suffixes to legitimate words—is irritating and distracting; it makes writing that did not need to be technical sound stiff and imposing. Look for cases like these in your first drafts; do not write

- *non-English speaking* when you mean *unable to speak English*
- *attendee* when you mean *participant*
- *interviewee* when you mean *respondent*
- *systematize* when you mean *arrange* or *order*

This is not to say that all words created by adding an "ee" or an "ize" are incompetent. *Cannibalize* is a marvelous word that is quite useful in engineering and manufacturing, and for which there is no replacement.

Even Shakespeare, in *King Richard the Second,* uses the word *monarchize* —although I suspect he was being sarcastic.

Also, be careful of misappropriating technical terms for use in inappropriate places. The words *maximize, minimize,* and *optimize* are essential terms in management science. But they should not be used loosely as synonyms for *increase, decrease,* and *improve.* (Use these terms only when talking about mathematical functions with definable limits. You can neither "maximize" a company's sales, for instance, nor "optimize" a trainee's learning—unless there is some mathematical function that describes the performance limits for those processes.)

BEFORE:	This approach will *maximize* our sales.
AFTER:	This approach will *increase* our sales.
	or
AFTER:	This approach will *maximize* our market penetration.
BEFORE:	The draft had *minimal* errors.
AFTER:	The draft had *very few* errors.
BEFORE:	Try to *optimize* this program.
AFTER:	Try to *reduce the cost* of running this program.
BEFORE:	What is the *optimum* approach to the OPEC cartel?
AFTER:	What is the *best* approach to the OPEC cartel?

Do not abuse terms from communications and computing. Do not call a favorable reaction from an audience "positive feedback" unless you want to sound like a pretentious fool. Do not refer to movie projectors as "hardware" and movies as "software" unless you are with other people who talk in this strange way. (If you are in computers, be careful what you say in front of your children, lest they grow up thinking that "transparent" means "invisible.")

Being Imprecise

Showing off—using bigger words than you need to—is bad on principle. And it is also bad for practical reasons.

Fancy words, more often than not, have a meaning that differs slightly from that of their simpler substitutes. So, when I tell you to prefer the plain to the fancy I am also telling you to choose the *best,* most *precise* word—which will not necessarily always be the plainer of the two.

The key to choosing the right words is knowing precisely what words mean. To treat words that are roughly equivalent as synonyms can be a tragic error, the same error that the thesaurus makes on every page.

(The thesaurus, or dictionary of synonyms, is a treacherous book. Beloved as it is by college students trying to make their writing sound fancier, the thesaurus casually clusters words of vague similarity under the heading of synonyms. If you use a thesaurus without a good dictionary at your side, you will probably write something barbaric.)

I tell my clients to *presume* that there are no exact synonyms. I do *not* tell them to *assume* that there are no exact synonyms. *Presume* and *assume* are a good example of a pair of words that are used interchangeably, even though they are not synonyms at all. To *presume* something is to treat it as true, whether or not it is true and whether or not you believe it. (In a typical criminal trial, for example, hardly anyone in the room believes the defendant is not guilty; yet the court does *presume* him not guilty.) In contrast, to *assume* something is to take it as true without proof or evidence.

Thus, the two terms are conspicuously different in meaning. (My thesaurus calls them synonyms.) If I had told you to *assume* that there are no synonyms in English, I would have told you to believe something of questionable truth. On the other hand, by telling you to *presume* it, I am merely urging you to look skeptically at terms you think are synonymous.

Consider these terms: *keep, preserve, retain, maintain.* All these terms are "synonyms," and all, in turn, are synonymous with the word *save.*

Why choose one word over another? Because the last three terms have special meanings:

- If you *preserve* something, that something is perishable.
- If you *retain* something, that something is likely to get lost or escape or break out of your possession.
- If you *maintain* something, that something is likely to run down or break down unless cared for.

These are not the only meanings of these terms. (Of course, most words have more than one meaning.) But they are the essential meanings, and, moreover, you already knew these meanings without looking them up! Amazingly, the most straightforward way to find the differences in meaning between similar terms is simply to pause and think of cases in which you would use one but not the other. Enough examples will form a pattern, and the meaning, therefore, is derived by finding an explanation for the pattern. (It's a lot like fitting a curve to a scatter of points.)

If you do not want to communicate any of these specialized meanings, by the way, then *save* will do.

This habit of writing in fancy words can actually *distort* your meanings. The word *indicate,* for example, means *to point.* If you are in the habit of saying "the data indicate" when all you mean is that "the data show," you are not just showing off but saying that the data are more "pointed" than they actually are. (Some people seem to think that *indicate* means *intimate.*)

The most serious lack of precision, therefore, is in using fancy words without regard to their special meanings. But there are other problems too.

Do not confuse *formality* with *vagueness.* Use the most specific and precise language you can. For the most part, writers are vague when they are unsure of their facts or trying to hide something. If you are vague all the time, your reader will find you devious.

Among the thousands of typical examples, here are a few:

TOO VAGUE	BETTER
financial resources	money
unfavorable weather conditions	rain, snow
equipment malfunction	broken printer
human relations skills	tact
human resource development	training
support personnel	clerks
analytical model	formula
effective communications	clear writing

Your aim is to be as precise and limited as you need to be: to find the most relevant subset of items from the larger, vaguer set.

Also resist the practice—popular with computer people, sociologists, and police officers—of assigning everyday objects and places new names. Resist the temptation to give an air of science to everyday things. (Remember TV's Coneheads, the aliens who refer to eggs as "chicken embryos" and families as "family units.")

INSTEAD OF	WRITE
processing unit	processor
printing unit	printer
transportation equipment	trucks
employment facility	office building
secretarial station	desk
home environment	house
client environment	market
photographic equipment	cameras and lights
telephonic unit	telephone
audio playback output device	loudspeaker
linguistic characters	letters
banking facility	bank
retail facility	store
retail fuel outlet	gas station
postal center	post office

One of the most serious consequences of vagueness and showing off is that words with strong, special meanings gradually lose their meanings—and need qualifiers to restore their original impact. If, for example, we had not begun to use *committed* in such a light, careless way, we would not need to write *totally committed* to make our point.

Look at these common examples:

EXCESS QUALIFICATION	ORIGINAL MEANING
completely devoted	devoted
utterly rejected	rejected
utterly unique	unique
perfectly clear	clear
completely compatible	compatible
completely accurate	accurate
quite precise	precise
quite innovative	innovative
radically new	new

The only time to qualify these terms is when they are used in a special or odd sense. Otherwise the qualifiers you want are already "wired in" to the words themselves. (Be careful, though. There is no such thing as *slightly devoted* or *somewhat unique* or *almost clear* or *relatively precise.* Use such terms with caution.)

Before moving to the next topic, I am obliged to say that there is also writing that is *too precise.* One of the paradoxes of technical writing is that, believe it or not, too much precision can make your writing *unclear.*

The degree of precision, as in other things, should be determined by the needs of the occasion: the audience and purpose. For some readers the details, qualifications, exceptions, contingencies, and qualifiers are irrelevant. But the reader, who may not know this, studies the exceptions and tributary items rather than the main point of the passage.

Be cautious of too much detail. Especially when you are answering a question or following an instruction, read carefully to be sure you have not exceeded the charge.

Avoid technical distinctions that make no difference to your reader. Do not write *upward compatible* when *downward compatible* is beside the point; write *compatible.* Do not insist with scientific elegance that you have a *sample drawn from a representative sampling plan* when all your reader needs is *reliable sample.* Do not put *sub* in front of things (subpopulation, subgroup, subsample, subset) unless the prefix is needed to make a point.

And refrain from putting all your exceptions and contingencies before the general claim.

Instead of

> Given the small sample, unless we are willing to tolerate huge sampling errors and mostly empty cells in the matrix (or unless we are willing to tolerate a nonparametric

form of contingency-table analysis, which would also probably be unacceptably inexact), none of the parametric techniques for multivariate analysis (especially analysis of variance and covariance and canonical correlation—for which we have already developed processing software) is entirely feasible.

write

Given the small sample, multivariate statistical analysis is not feasible. If we did analysis of variance or covariance, or canonical correlation, the sampling errors would be unacceptably large and most of the cells would be empty. Even a nonparametric, contingency-table analysis would be too inexact.

Wasting words

The most common complaint among writers and readers of technical communication is "wordiness." Fortunately, this is the easiest problem to solve.

Most wordiness comes from two sources: long-winded functional phrases and defective verbs. I talk about verbs in the next chapter, but now I shall show you the phrases that account for 70 or 80 percent of the excess verbiage in your first drafts.

General Purpose Throat-Clearers When speakers need a moment to find a word or phrase, they clear their throats (or use some other ploy). When writers are stuck, they use characteristic phrases and expressions: throat-clearers. These should be cut from the first draft.

Be wary of such expressions as

in order to	the nature of
in order for	the case of
the reason for	relative to

BEFORE:	*In order to* explain *the reason for* this proposal, we must first explain our previous activities *relative to* the inventory control problem.
AFTER:	To explain why we wrote this proposal, we must first explain our previous work on the inventory control problems.

Space- and Time-Wasters When we write about intervals and dimensions, we usually use too many words. Express measurements and units as simply as possible.

INSTEAD OF	WRITE
in length	long
in height	high
10 square feet of space (or area)	10 square feet
3 cubic yards of volume	3 cubic yards
distance of 4 kilometers away	4 kilometers away
in a westward direction	westward
rectangular in shape	rectangular
period of time	time, or period
twenty-months' duration	twenty months
interval of time	interval
three hours long	three hours
during periods of	during
three hours of time	three hours

Avoid such terms as *calendar month* and *calendar year.* Write just *month* or *year* —unless the word *calendar* is necessary to distinguish your statement from *person-month* or *person-year.*

Fear of "About" Too many writers seem unwilling to use the clear, plain *about* when something longer or fancier is available.

The word *about* has two main meanings:

- approximately
- on the subject of

ABOUT I: Write *about* instead of

approximately	within the ballpark of
in the vicinity of	more or less
on the order of	in the range of

ABOUT II: Write *about* instead of

relative to	in connection with
with regard to	with respect to
regarding	pursuant to
re	reference
on the subject of	in relation to
relating to the subject of	respecting
in the matter of	in reference to

You may use these longer equivalents, sometimes, to avoid tiring repetition or to make a finer point. But, in general, scratch them out of your first drafts.

Fear of "Use" The word *utilize* is an unnecessary substitute for *use.* Stop writing *utilize.* Use *employ* if you want to avoid the phrase *using people.* But generally write *use* instead of

- utilize
- put to use
- employ
- make use of
- make utilization of

BEFORE:	The company *utilizes* econometric models in its market forecasts.
AFTER:	The company *uses* econometric models . . .
BEFORE:	Our clerical staff is *underutilized.*
AFTER:	Our clerical staff is *underworked.*
BEFORE:	We proposed the *utilization* of a stricter security system.
AFTER:	We proposed the *use* of a stricter security system.
	Or
AFTER:	We proposed a stricter security system.
	Or
AFTER:	We urged stricter security.

Various dictionaries offer special, distinct meanings for *utilize.* One says it means "put to profitable use"; another, "use for some specific purpose"; and a third, "use for some unusual or unexpected purpose." No one knows for sure if the word *utilize* means anything at all! Stop using it. Please!

Fear of "Now" and "Then" Too many writers substitute fancy or wordy expressions in place of the clear *now* or *then.*

Write NOW instead of	Write THEN instead of
at this time	at that time
at this point	at that point
at this point in time	at that point in time
after further consideration	at that date
upon further reflection	as of that time
at present	up until that time
at the present time	during the past
as of this time	in that period
our current belief is	during that earlier time frame

Write NOW instead of	Write THEN instead of
as of today	within those earlier parameters
so far as we can see at this point in time	in the past

To avoid overuse of *now,* write *today, lately,* or *recently.* To avoid overuse of *then,* write *earlier, later,* or *before.* (Never use *presently* for *now:* It means *soon.*)

Fear of "To" Do not yield to the habit of embellishing the word *to* with unnecessary words.

WRITE *TO* INSTEAD OF

- in order to
- so as to be able to
- with a view to
- as a way to
- for the purpose of being able to
- as a means to

In most cases, the word *to* all alone will say what you mean. Of course, in cases where these additional words carry a special meaning, leave them in.

Fear of "So" *So* is a clear, respectable word. Do not be afraid to use it in either of its two senses: as a synonym for *in order that* or as a simple way of saying *thus.*

SO—I (in order that; so as to be able; for the reason that)

BEFORE:	We revised the manual *in order that* the operators would need fewer passes at the final reports.
AFTER:	We revised the manual *so that* the operators would need fewer passes . . .
BEFORE:	*So as to be better able* to detect unwarranted machine use, we added additional security features.
AFTER:	*So that we can* better detect unwarranted machine use . . .

SO—II (thus, therefore, consequently, as a result)

BEFORE:	*Thus,* our conclusion is that . . .
AFTER:	*So,* we conclude that . . .

SO—II (thus, therefore, consequently, as a result)

BEFORE:	*Therefore,* we cannot afford an additional console this year.
AFTER:	*So,* we cannot afford an additional console this year.

Fear of "By" and "With" Do not be afraid of *by* and *with.* Use them instead of longer, clumsier expressions.

WRITE *BY* OR *WITH* INSTEAD OF

- by means of
- by using
- utilizing
- through the use of
- via
- by employing
- involving

BEFORE:	Clear the data base *by means of* a ZERO command.
AFTER:	Clear the data base *with* a ZERO command.
BEFORE:	Excessive delays in the changeover can be prevented *through the use of* early training.
AFTER:	Excessive delays in the changeover can be prevented *by* early training.
BEFORE:	*Utilizing* a different file structure, this report could be run with much less CPU (Central Processing Unit) time.
AFTER:	*With* a different file structure, this report could be run with . . .

Fear of "For" Many writers overlook *for* and use such longer expressions as *in order for* or *for the purpose of.* Further, it is often possible to use a simple *for* in place of such verbs as *to accomplish, to ensure, to allow, to achieve,* and *to effectuate.*

FOR—I

BEFORE:	*In order for* this report to be produced, you will need to link three files.
AFTER:	*For* this report you will need to link three files.

	FOR—I (continued)
BEFORE:	They upgraded their cash registers *for the purpose of* better sales and inventory data.
AFTER:	They upgraded their cash registers *for* better sales and inventory data.
	FOR—II
BEFORE:	You usually need only three commands *to accomplish rotation* of the drawing.
AFTER:	You usually need only three commands *for rotation* of the drawing.
BEFORE:	*To ensure* security, we use a hierarchy of passwords.
AFTER:	*For* security, we use a hierarchy of passwords.

Fear of "If" Use the word *if* in place of such ponderous expressions as *should it prove to be the case that* or *in the event that.*

BEFORE:	*In the event that* there are blanks in one of the columns, display the arithmetic for that column.
AFTER:	*If* there are blanks in one of the columns, display the arithmetic . . .
BEFORE:	*In those cases when* the report is perfect, send it to the remote printer.
AFTER:	*If* the report is perfect, send it to the remote printer.
BEFORE:	*On those occasions when* scientists are the main users of the system, FORTRAN is the best language.
AFTER:	*If* scientists are the main users of the system, FORTRAN is the best language.

When and *where* are also acceptable substitutes in many of these cases.

Fear of "Of" Avoid ponderous, complicated expressions that can simply be replaced with *of.*

WRITE *OF* INSTEAD OF

- derived from
- realized from
- noted in
- associated with

- inherent to
- obtained from

BEFORE:	Notice the cost advantage *derived from* using readers instead of typewriter inputs.
AFTER:	Notice the cost advantage *of* using readers instead of . . .
BEFORE:	We decided to study the impact *realized from* our magazine advertisements.
AFTER:	We decided to study the impact *of* our magazine advertisements.
BEFORE:	This report generator is free from the formating problems *noted in* our first version.
AFTER:	This report generator is free from the formating problems *of* our first version.

Double-Talk Try not to repeat yourself, and do not say the same thing twice! Certain phrases are redundant and should be trimmed.

TOO MUCH	MUCH BETTER
integral part	part
repeat again	repeat
active consideration	consideration
refer back	refer
past history	past or history
present status	status
true facts	truth or facts
midway between	midway or between
entirely complete	complete
in ten years from now	in ten years

If you know the meanings of the words in these phrases, you know they are redundant. There is no such thing as inactive consideration, so *active consideration* must tell us too much. Further, all stops are "complete stops": as in driving, "rolling stops" do not count.

Some redundant phrases explain too much, defining terms that are already clear:

TOO MUCH	MUCH BETTER
unknown stranger	stranger
tangible enough to be felt	tangible
consensus of opinion	consensus
a person-day of effort	a person-day
baffling enigma	enigma
visible to the eye	visible
disqualified for ineligibility	disqualified or ineligible
concatenated together	concatenated (linked)

The Present/Provide Trap Most of the times we write the words *present* or *provide* (and their associated forms), we really mean something far simpler than *presenting* or *providing.* (*Furnish* sometimes gives us the same problem!)

BEFORE:	Details are *provided in* Chapter 4.
AFTER:	Details are *in* Chapter 4.
BEFORE:	The purpose of this chapter is *to present* a series of typical errors and error messages.
AFTER:	The purpose of this chapter is *to show* a series . . .
BEFORE:	The transition *furnished us with* several problems.
AFTER:	The transition *had* several problems.
BEFORE:	Our goal *is the provision of* timely management information.
AFTER:	Our goal *is* timely management information.

(Do not use *provision* as a substitute for *providing;* do not use *presentation* as a synonym for *presenting.*)

PHRASES

Now that you have replaced your fancy and imprecise vocabulary, and substituted simple words for long-winded expressions, you should review your use of phrases. In particular, ask

- Are my phrases in the right places?
- Do my introductory phrases dangle?

- Can my phrases be compressed?
- Are my phrases too compressed or compact?
- Are my phrases trite and tired?

Solving these problems is a bit more difficult than replacing inappropriate words. Sometimes, for example, you have to move a phrase from one part of a sentence to another. But after a few hours of practice, this too becomes easy.

Misplaced Phrases

Descriptive phrases (particularly the prepositional phrases beginning usually with *on, by, with, of, in, from*) should be close to the words they describe or modify. Typically they will come after, occasionally before.

Because the "normal" expectation is that such phrases modify the words before them, a careless misplacement of the phrase (natural in the first draft) will confuse and distract the reader. Be sure your descriptive phrases cling to the terms they describe. The greater the separation, the greater the chance of ambiguity or misreading. Usually, the best location for the phrase is right after the term it describes.

BEFORE:	I have the resumés for four promising analysts *on my desk.* (Why are these analysts on your desk?)
AFTER:	I have *on my desk* the resumés for four promising analysts.
BEFORE:	I wrote my evaluation of the new CRT *in the quarterly report.* (Is there a CRT in the new quarterly report?)
AFTER:	I wrote *in the quarterly report* my evaluation of . . .
	Or
AFTER:	*In the quarterly report* I wrote my evaluation . . .
BEFORE:	Notify the manager of breakdowns *in accordance with department rules.* (Only those breakdowns?)
AFTER:	*In accordance with department rules,* notify the manager of breakdowns.

If you are careless, you might write something barbaric:

> The car was designed by our new safety engineer with an impact-absorbing rear end.

Also be especially careful of putting two of these descriptive phrases in a row—especially if they both begin with the same word. There is nothing illegal about such an arrangement, but it should be used only when the second phrase is meant to modify the last word in the first phrase (or sometimes the last phrase as a whole). Look at these examples:

BEFORE:	He sketched the new system *on* the backs of several envelopes *on* the way to the office.
AFTER:	On the way to the office, he sketched the new system on the backs of several envelopes.
BEFORE:	Sit comfortably at the console *with* the manual *with* the red cover close by.
AFTER:	Sit comfortably at the console with the red-covered manual at your side.
	Or
AFTER:	With the red-covered manual at your side, sit comfortably at the console.
BEFORE:	The formulas for estimating inflation are *in* the ECON file *in* the FINC data base.
AFTER:	The formulas for estimating inflation are in the ECON file *of* the FINC data base.

Even when these "stacked" descriptive phrases are clear (as in the last example), they still sound peculiar. Change them, at least enough to eliminate the repeated preposition.

Danglers

Danglers, including the infamous "dangling participle," terrorize many engineers and scientists. Fearful of making the error but not quite sure what the rules are, some writers avoid *all* introductory phrases. This frees their writing from introductory danglers, but it also tends to create boring, inside-out sentences.

The rule on danglers is simple. *Introductory phrases must be tied in meaning to the subject of the sentence. And the subject of the sentence must appear immediately after the comma that sets off the introductory phrase.*

If the sentence begins: "Trying to contain the costs of plant maintenance . . ." then the next word in the sentence must be whoever or whatever is *trying to contain* those costs:

> Trying to contain the costs of plant maintenance, *the vice-president for operations* approved the new roofing plan.

Any other word or phrase after that comma would be a disastrous dangling participle, almost as bad as the famous

> After eating my lunch, the director of marketing told me how much he valued my patent.

In this last sentence, there is a dangling participle—*unless you want to say that the director of marketing ate your lunch.*

Use good sense in all your introductory phrases. Do not be fainthearted, like the businessman I know who instructs his secretaries to recast all sentences that have a word ending in *-ing* in the opening phrase! Study these examples until you get the point!

Introductory Phrase

BEFORE:	As an industrial hygienist, my schedule is filled with travel.
AFTER:	As an industrial hygienist, *I* have a full travel schedule.
BEFORE:	With more than five hundred employees engaged in research and development, there is no aspect of modern computer science that is not in the company's expertise.
AFTER:	With more than five hundred employees engaged in research and development, *the company* has expertise in every phase of computer science.
BEFORE:	As an example of carelessness, the manager cited the report of the maintenance contractor.
AFTER:	As an example of carelessness, *the report* of the maintenance contractor is worth noting.
	Or
AFTER:	The report of the maintenance contractor, according to the manager, is an example of carelessness.

Introductory Participle Often, the opening phrase of a sentence contains a participial form of the verb, usually the active participle ending in *-ing.* Whenever there is a participle in the introductory phrase, the subject of the action in that participle *must* be the subject of the sentence and must appear immediately after the comma used to set off the phrase.

BEFORE:	Wishing to analyze the failures of the project, a meeting was held between the sponsor and the contractor.
AFTER:	Wishing to analyze the failures of the project, *the sponsor and contractor* met.

BEFORE:	When initializing the system with a coldstart, a tape containing the operating system software is loaded.
AFTER:	When initializing the system with a coldstart, *the operator* loads a tape containing the operating software.
	Or
AFTER:	When initializing the system with a coldstart, (you) load a tape containing the operating software.

There is also a dangling *past* participle—usually a word ending in *-en* or *-ed.*

BEFORE:	Located in the small intestine you will find the villi.
AFTER:	Located in the small intestine, the villi can . . .

By the way, participles can also dangle at the end of a sentence. The rule is simple: When an *-ing* verbal starts a phrase at the end of a sentence, the subject of the verbal should be the noun that preceded it.

Be careful of

The patient left the hospital urinating freely.

Introductory Infinitive Like participial phrases, infinitive phrases must also refer to the subject of the sentence, which should appear right after the comma.

BEFORE:	To enlarge their market, retail showrooms were opened in office centers.
AFTER:	To enlarge their market, *they* opened retail showrooms in office centers.
BEFORE:	To ensure security of the company's confidential data, a hierarchy of passwords is used.
AFTER:	To ensure security of its confidential data, *the company* uses a hierarchy of passwords.
BEFORE:	To revise the row definitions, DROX is entered.
AFTER:	To revise the row definitions, *(you)* enter DROX.
BEFORE:	To find the way to the infirmary, red flags have been posted on the shortest path.
AFTER:	To find the way to the informary, *you* should follow the red flags posted along the shortest path.

Wordiness, Again

Earlier in this chapter I listed scores of phrases that could be converted into single words. Sometimes you can also improve a sentence by converting a phrase of several words into a simple suffix. The suffixes most useful in this case are *-less, -like, -ly, -ing,* and others.

BEFORE:	The tests were run without a flaw.
AFTER:	The tests were flaw*less*.
BEFORE:	The microfiche are transported in a device that resembles a small elevator.
AFTER:	The microfiche are transported in a small, elevator-*like* device.
BEFORE:	The error message describes what went wrong in a very explicit fashion.
AFTER:	The error message describes explicit*ly* what went wrong.

Overcompression

When I urge people to collapse phrases into words and suffixes, I sometimes worry that they will go too far.

Although the problem of wordiness is far more prevalent than the problem of overcompression or overefficiency, there are, nevertheless, too many people who try to jam too much information into a single phrase.

One of the most typical bugs in technical writing, expecially among computer people, is the *noun string* (that is, the string of nouns).

No writer can be understood if he or she writes in strings of nouns, that is, two or more nouns in a row. Even a two-word string like *executive decision* could mean

- a decision available to executives, or
- one of several ways to lead, or
- the choice of having executives or not.

INSTEAD OF	WRITE
problem responsibility changes	changes in the assignment of responsibilities
graphics construction language	language for constructing graphics
component reference designators	designators for referring to components

for your information	under separate cover
hand you herewith	up to this writing
hereby advise	we remain
herewith enclose	we trust
in conclusion would state	whereas
in connection therewith	with reference to (relative to)
in due course	would advise, would appear, etc.
in re	

I know what you are thinking. Without these stock phrases it will be impossible to write a memo or letter. Nonsense! The sooner you change *enclosed please find* to *here is,* the better for all of us.

Stewardess Talk Be especially wary of what I call "stewardess talk," standard phrases of politeness. For example:

- additional assistance
- do not hesitate
- regret any inconvenience
- we ask you
- it's been a pleasure
- anything in our power
- comfort and convenience
- we trust that
- certainly hope
- cordially invite
- pleasant experience (avoid all phrases with the verb *experiencing,* such as *experiencing difficulties*)

Do not use any phrases you associate with rehearsed speeches of friendship and accommodation. Although intended to be sincere and warm, this language has the opposite effect: false and cold.

Figurative Expressions Unless you are a very good writer, avoid two figures of speech: metaphors and similes. Be especially careful of "mixed metaphors," two or more figures expressed in the same phrase or sentence. For example:

- Let's not fuel the fire of galloping inflation.
- We backed him to the hilt when the chips were down.
- The shadow of inflation hovers over the economy.
- Discard any ideas that don't ring bells.
- All hopes of a solution flew out the window.
- They march to the beat of a different agenda.

A HOPEFUL WORD

Sometimes, when I finish talking to a group of engineers and scientists about words and phrases, I notice that they have a bruised look about them. As though I had been slapping them with insults and criticism, as though I had been calling them phonies, showoffs, and bores.

I do not want you to be hurt or intimidated by all this criticism. And do not be defensive, either. You never *decided* to write the way you write. You did not choose to write *utilize* or *pursuant to* or *in the event that.* These are just habits you picked up from the gang at school and work. (Sadly, you picked up several of them from your teachers.)

Listen. No one is attacking your intelligence or education. I am talking about *editing,* not *writing.* These scores of mistakes and flaws are typical of everyone's first drafts, including mine, including those of writers who are much better than you or I.

As I have said several times, especially in Chapter 8 ("Attacking the Draft"), it is wrong to worry about any of these matters while you are writing a first draft. I do not care if your first draft contains every blunder listed in this chapter. I *do* care, though, if these first-draft blunders actually leave your desk and get to your reader. I worry about your career. I worry about the well-being of the language.

Again, this chapter contains the *easiest improvements to make in your first draft.* It does not contain advice on how sentences should come out the first time. When you have finished with this chapter, I want you to reach two conclusions:

- Anyone who does not edit every sentence in a first draft is bound to send out pompous, wordy, unclear, uninteresting material.
- Anyone who wants to master a few simple tricks and replacement rules—and take very few minutes to apply them to his or her texts—can make dramatic improvements in his or her writing.

This is the easiest part. Now the bad news. What comes next is harder—although nowhere near as hard as organic chemistry or second-semester calculus. Many of you will go no further than this with your editing. For that much, at least, your readers will be grateful.

But to become an effective writer you must also study the architecture of sentences and paragraphs. You must ponder some problems of (dare I say it?) grammar and syntax.

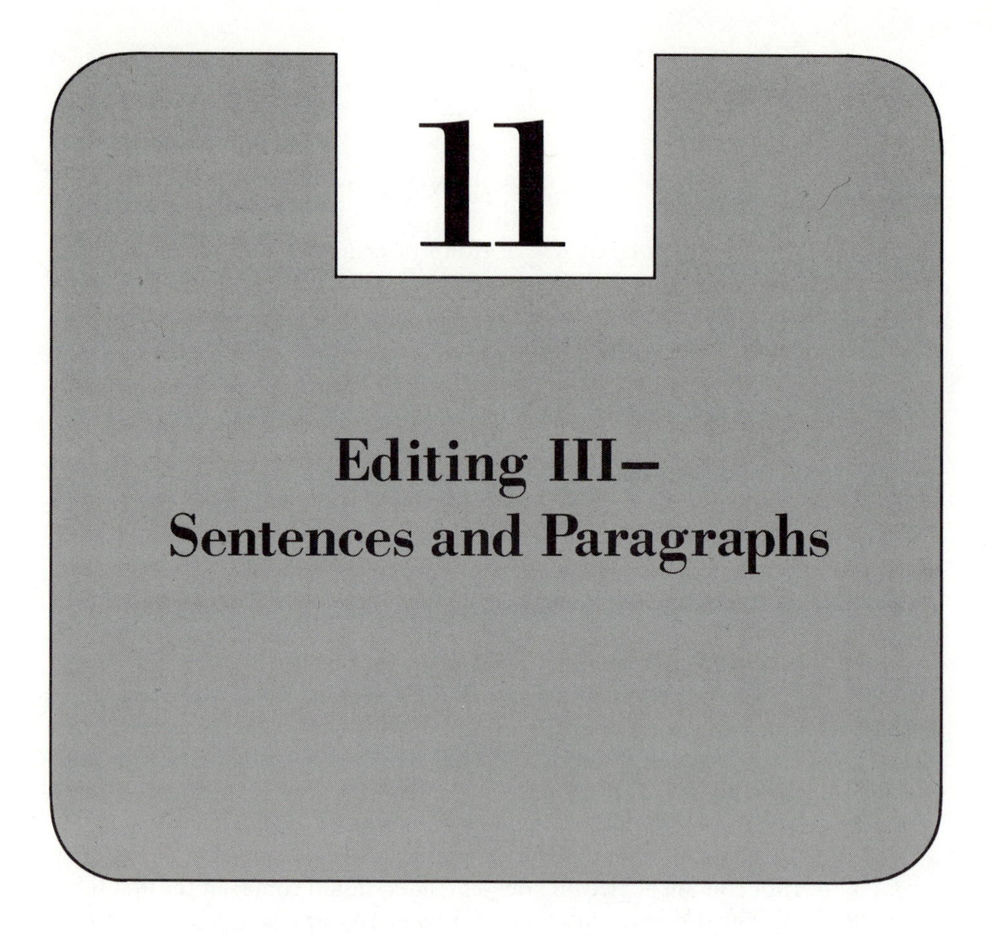

11

Editing III—Sentences and Paragraphs

AFTER WORDS AND PHRASES, THEN WHAT?

By now you have solved the simplest editorial problems. You have replaced *utilize* with *use, in order that* with *to, prioritize* with *rank,* and so forth. You have finished most of the "easy" editing, the changes that take little more than simple substitutions from replacement tables.

Even though it has been easy, do not think it has been unimportant. If all you ever do is make the changes described in the preceding chapter, you will still be a more effective writer than you were—and probably better than many of your colleagues.

But do not become too confident. There is still much basic and important editing to go, most of it requiring you to do a bit more than simple replacements. Hereafter you will have to *think,* possibly experiment with alternatives. You are moving further away from what is clearly right or wrong and more into the area known as "style"—the craft of making your writing more accessible and interesting.

Most of the remaining work involves "tuning up" your sentences and paragraphs, making them work better. The general rule for paragraphs and sentences is the same: *They should begin well and grow increasingly more interesting.*

This advice may come as a shock to you. Until now you may have thought that the most interesting part of a sentence should be the beginning, or that the most important part of a paragraph should be its opening sentence. (You may even think that you read something like that earlier in this book.) But you are wrong!

True, everything you write, and every part of everything you write, must begin well, in a way that captures the attention. But the effective writer does more than capture the reader: *The effective writer draws the reader in deeper and deeper.*

The greatest weakness in the writing instruction given in schools and colleges is that teachers almost never tell students that the main information in a sentence should usually appear *at the end,* and that paragraphs should build in interest from the first sentence. It is hardly surprising that most writers compose sentences that fall off at the end and write paragraphs that go nowhere after the first sentence. (It is as though people were writing just for the "speed readers," who tend to read only the beginnings of sentences and the first sentences of paragraphs.)

Although writing masterly sentences and paragraphs is a craft that takes years to develop, you *can* quickly learn the most typical errors that make the average sentence less readable than it should be and that make the average paragraph choppy and unsatisfying.

HOW TO EDIT YOUR SENTENCES

To improve your first-draft sentences, follow these rules:

- Put the new and interesting stuff at the end
- Trim away extra words
- Write with verbs instead of nouns
- Write in the active voice whenever you can
- Put words, phrases, and clauses in parallel
- Emphasize and contrast your points
- Do not be too compressed or complicated
- Do not be too childlike and simplistic
- Vary your sentence patterns

Put the New and Interesting Stuff at the End

You have choices to make. You are in control. What happens in your sentences is up to you. You are the designer.

Consider these two sentences:

A. A physician's main job is to reduce pain and suffering.
B. Reducing pain and suffering is the main job of a physician.

Which of these sentences is "correct"? Both are understandable; both are free from errors of grammar or style.

Does it matter which is "correct"? Could you use them interchangeably? Do you have any idea of what criteria you might use to choose one or the other?

The answer is simple. The correct version depends on what point you are making (and what points you have already made). Consider these two paragraphs:

A. Physicians have many responsibilities. They perform research and teach other physicians and sometimes help to prevent disease. *But a physician's main job is to reduce pain and suffering.*
B. Most of us see pain and suffering only once in a while. A teacher will occasionally have a very sick student. A carpenter will, two or three times in his career, help an injured co-worker. Even police and firefighters go for days at a time without seeing a person in distress. *But reducing pain and suffering is the main daily job of the physician.*

In each of these two paragraphs, the italicized version of the sentence is correct because it carries the reader's attention from what is old to what is new, from what has been established to what is to be proved, from the given to the hypothetical, from the background to the foreground, from the premise to the conclusion, from the subsidiary to the main.

Consider this pair:

A. Our new word processor has improved the quality of our letters and sales literature. *The main effect of the word processor, though, has been improved secretarial morale.*
B. Our secretaries have received a new benefits package, including more vacation and more training opportunities. *But the improved secretarial morale has mainly been the effect of our new word processor.*

Again, notice the sequence in the emphasized sentences, the flow of information from old to new. No matter what you may think, and no matter what you may have been told, the new, important, interesting information in a sentence is supposed to appear *at the end.* When we are composing first drafts, though, we tend to blurt out the interesting material at the beginning.

Think of a sentence as a story that builds suspense and reaches a climax. Put the surprise at the end.

BEFORE:	The end of a sentence is the place where the interesting information should appear.
AFTER:	The place in a sentence where the interesting information should appear is the end.

BEFORE:	The detailed analysis of trends and how they affect the business of the company is the purpose of the data base.
AFTER:	The purpose of the data base is to analyze trends and how they affect the business of the company.

Although you could imagine cases where the "Before" versions might be preferred, you are much safer with the "after" versions.

Moreover, putting sentences into proper sequence not only makes the sentence itself more readable but also prepares the way for the next sentence. Ordinarily, the new, main information in the first sentence becomes the old, established subject of the next sentence:

Clumsy: Harvard Business School is among the main attainments of our new director. He studied there in 1976–77.

Better: Among the main attainments of our new director is an M.B.A. degree from the Harvard Business School. He studied there in 1976–77.

Notice the easier transition across the sentences—an issue that will come up again when I talk about paragraphs.

The beginning of a sentence—after introductory phrases and clauses—is usually the subject, and the end is usually the predicate. Predicates contain the verb and the object of the verb (if there is an object) or some complement to the verb. Grammar aside, what you must know is that the "action" is in the predicate, not the subject. Yet, not only do many writers put the main information into the subject instead of the predicate, they even try to throw the verb itself into the subject noun!

When we are composing, we often jam all the interesting information into the subject of the sentence and leave nothing for the predicate. Although the following sentences are grammatically correct (they *do* have a predicate), they are awkward and dull. Such predicates as *exists, is provided,* and *is to be noted* are signs of the problem.

NO:	Printing of user directories showing creation date and time *is provided.*
YES:	The system *prints* user directories that show the date and time of creation.
NO:	The possibility of producing the product locally *is to be noted.*
YES:	We can *produce* the product locally.
	Or
YES:	Perhaps we can *produce* the product locally.

NO:	The need for better computer equipment in the high schools *exists.*
YES:	The high schools *need* better computer equipment.

Sentences with empty predicates often have subject nouns ending in the syllables *-tion, -ment,* or *-ing.* Use this as a test of your own first drafts.

This problem of sequence, that is, the need to put the main information at the end, is most evident in *complex sentences.* If you recall your grammar, you will remember that complex sentences are those that contain one or more *dependent* or *subordinate* clauses. These clauses—called dependent because they cannot stand alone as sentences—are easily spotted because they typically start with such words as *if, so, when, although, because,* and *since.* (These are not all the subordinating words; nor does every dependent clause begin with a special word.)

One of the few continuing conflicts between editors in general, and technical and business editors in particular, is that most technical editors tend to put the main clause at the beginning, *thereby ensuring that the sentence will fade in interest.* The general editor, however, knows that subordinate and dependent clauses should usually (that is, 70 or 80 percent of the time) be at the beginning of the sentence, thereby saving the main clause (with the new information) for last.

This problem is exacerbated by the commonplace fear of starting a sentence with the word *because,* a groundless phobia that encourages people to put the *because* clause last instead of first.

NO:	We have stopped using the U.S. mails *because* of the delays we described.
YES:	*Because* of the delays we described, we have stopped using the U.S. mails.
NO:	This plan will never satisfy the sponsor, *even though* the consultants love it.
YES:	*Even though* the consultants love this plan, it will never satisfy the sponsor.

To further complicate things, some writers happen to sense that the main information in the sentence should be at the end, but because they are in the habit of putting the dependent clause at the end, they put the main information into the dependent clause (instead of the independent)—at the end of the sentence!

NO:	The reactions from the meetings were positive, so implementation of the design began.
YES:	Because the reactions from the meetings were positive, we began the implementation of the design.

Trim Away Extra Words

Most of the extra words in your sentences should have been eliminated by following the rules in the preceding chapter. There are a few more word-saving devices to go, though.

First, look for opportunities to convert **clauses into phrases.** For example:

NO:	*While they were waiting for the report to print,* they reviewed the last run.
YES:	*Waiting for the report to print,* they . . .
	Or
YES:	*During the wait for the report,* they . . .
NO:	*Once the meeting had been finished,* we were ready to draft the specs.
YES:	*After the meeting,* we were ready . . .
	Or
YES:	*The meeting being over,* we were ready . . .
NO:	We incorporated the changes *that were demanded by the sponsor.*
YES:	We incorporated the changes *demanded by the sponsor.*

Next, look for those few cases where **whole clauses can be reduced to just a few words.** For example:

NO:	We were looking for a vendor *who had experience.*
YES:	We were looking for *an experienced* vendor.
NO:	Store these data in a file *to which all users have access.*
YES:	Store these data in a file *accessible to all users.*
NO:	He doubted that *the data were the most timely available.*
YES:	He doubted the *timeliness of the data.*

Be careful, though. Do NOT convert prepositional phrases into modifiers; do not replace *evaluation of the plan* with *plan evaluation.* That sort of trimming will make your sentences *harder* to understand.

Also, look out for **tautologies.** To a logician, a tautology is a kind of equation in which two statements are logical equivalents (A if and only if B). In editing, though, a tautology is a sentence that says the same thing twice. When you have written the same information twice, usually eliminate the *longer* term.

NO:	Train only qualified programmers who are suited to this job.
YES:	Train only qualified programmers.
NO:	We made a fatal tape error of the most serious kind.
YES:	We made a fatal tape error.
NO:	Here is a picture of our new and exciting JB-21, which we have just introduced.
YES:	Here is a picture of our new and exciting JB-21.

(Of course, if the longer term in the tautology has slightly more, or more precise, information, you may want to strike the shorter term instead.)

In general, feel free to cut any word or phrase that is irrelevant to your meaning or not useful to your reader. Do not, however, edit as though sentences were supposed to be logically elegant. As you will see later, good writers are more likely to say things twice, to give *extra* emphasis and contrast than to strip away every word that cannot be justified rigorously. If you must err, *err on the side of leaving words in.*

Write With Verbs Instead of Nouns

Technical people—especially in the computer and electronics industries—use too many nouns and not enough verbs. (Some editors call this "nominalization," but I cannot bring myself to call it that.)

A commonplace characteristic of the first draft is the so-called smothered verb, a potentially lively verb suffocating inside a dull, stodgy noun.

The most familiar smothered verbs are in phrases beginning with the words *have* or *make.*

SMOTHERED	LIVELY
have an objection	object
have knowledge	know
have reservations	doubt
have a suspicion	suspect
have concern	care, worry
make a distinction	distinguish
make a recommendation	recommend
make a suggestion	suggest
make a proposal	propose

NO:	I did not *have* a sufficient *knowledge* of the problem to *make a proposal* for a new system.
YES:	I did not *know* enough about the problem to *propose* a new system.

Smothered verbs can also be made in many other ways. Although the most common forms use *have* and *make,* there are also hundreds of expressions using *give, reach, do,* and others.

SMOTHERED	LIVELY
give an answer	answer
give an apology	apologize
give a justification	justify
reach a conclusion	conclude
reach a decision	decide
reach an end	end, finish
reach an agreement	agree
do an inspection	inspect, check
do a draft	draft
raise an objection	object
hold the opinion	believe
send an invitation	invite
hold a meeting	meet
furnish an explanation	explain
furnish a solution	solve
form a plan	plan

NO:	We *held a meeting* and *reached a decision* to *send* him *an invitation* to the bidders' conference.
YES:	We *met* and *decided* to *invite* him to the bidders' conference.

In writing done by technical people we also find an especially ornate and complicated form of smothered verb, using *accomplish, achieve, realize,* or even *effectuate.* To make things worse, this last group often appears in the passive voice of the verb.

SMOTHERED	LIVELY
separation was accomplished	they separated
a profit was realized	they profited
effectuate a system startup	start up the system
file linkage was achieved	the files were linked

NO:	The *calculations* to project interim costs *are accomplished* entirely in TREND.
YES:	TREND *calculates* the projections of interim costs.
	Or
YES:	TREND *projects* the interim costs.

As in every other case of replacement, the substitutions should be made with care. If the longer smothered form contains some subtlety or sound that is important to your meaning, use it. Otherwise use the verb on the right.

Technical writing is also filled with "weak servants," verbs of service that merely add extra words to otherwise clear sentences and smother the more lively verbs nearby. The most common offenders are *serve, use, conduct, perform,* and *carry out.*

	SERVE
NO:	This design *serves to explain* the three stages of the project.
YES:	This design *explains* the three stages of the project.
	USE
NO:	The electric pen *is used to correct* errors in the array on the screen.
YES:	The electric pen *corrects* errors in the array on the screen.
	CONDUCT
NO:	The agency will *conduct an investigation of* the purchase.
YES:	The agency will *investigate* the purchase.
	PERFORM
NO:	Comparison of the actual and projected *is performed within* the software.
YES:	The software *compares* the actual and projected.
	CARRY OUT
NO:	First, *carry out a search of* the literature.
YES:	First, *search* the literature.

And there are still other ways to smother, disguise, and suppress the verbs in a sentence. Almost any verb, if you work hard at it, can be hidden inside a *noun.* Often, when the apparent verb in the sentence is *to be (is, am, are, was, were, be),* there is a noun nearby that would make a more lively and interesting verb.

NO:	Their *decision* was to expand, rather than replace.
YES:	They *decided* to expand, rather than replace.
NO:	Their *approach* was a *separation* of the personnel and payroll files.
YES:	They *approached* it by *separating* the personnel and payroll files.
	Or
YES:	They *separated* the personnel and payroll files.

NO:	This technique is an *expansion* of the original security system.
YES:	This technique *expands* the original security system. (Or does the writer mean "just an expansion"?)

Verbs can also be disguised as *gerunds.* A gerund is a form of a verb that ends in *-ing* and acts as a noun. In the sentence "Running is good for you," *running* is a gerund and is the subject.

NO:	*Logging-in* is the next step after checking the acoustic coupler.
YES:	*Log in* after you check the acoustic coupler.
NO:	*Learning* the sign-on procedure should occur before *learning* the file-creation commands.
YES:	You should *learn to* sign on before you *learn to* create a file.

Of course, there is nothing wrong with gerunds. They are legitimate and often desirable. Merely be alert to cases—especially in instructions and commands—where the gerund is concealing the best verb.

Authors are endlessly ingenious in smothering their verbs. For example, the verbs *must, has to,* and *needs* can be concealed in a variety of nouns, adjectives, and even longer verbs. For example:

NO:	It *is necessary to* have complete documentation.
YES:	We *must* have complete documentation.
NO:	The operator on duty at 4:00 P.M. *has the responsibility for* logging off the system.
YES:	The operator on duty at 4:00 P.M. *has to* log off the system.
NO:	Updating these consumer loan accounts monthly *is a mandatory procedure.*
YES:	You *need to* update these consumer loan accounts monthly.

Write in the Active Voice Whenever You Can

In a misguided attempt to sound "impersonal," "detached," or "scientific," most technically trained people overuse the passive voice. The more normal voice is the active voice, in which the subject performs the action of the sentence. In the passive voice, the subject is *acted upon.*

Most engineers and scientists have heard some editor complain about the passive voice. But they are not quite sure what it is, and they are not entirely sure why it is so objectionable.

As you will see in the following examples, the main faults of the passive are that

- It often adds words to the sentence.
- It is harder to understand (a fact proved in tests).
- It invites style errors (for example, dangling participles).
- It conceals the identity of key people and actors.
- It sounds stuffy and phony.

Nevertheless, most writers, especially in the sciences, motivated by their fear of personal pronouns (never say "I") and conditioned by the bad habits of the entire industry, will invert or reverse the logical order of their sentences. Instead of saying "The company bought the package," they say "The package was bought by the company." In this process, they convert a clear, active verb to a less-clear passive verb in which the real actor or subject is hidden in a phrase beginning with the word *by.* For example:

NO:	A preventive maintenance manual *is left* on site *by* the field engineer.
YES:	The field engineer *leaves* a preventive maintenance manual.
NO:	Extensive editing *can be invoked* by the user.
YES:	The user *can invoke* extensive editing.
NO:	Flexibility *is not exhibited by* the procedure.
YES:	The procedure *is inflexible* (or rigid).

The only time you would deliberately choose this "by" passive form is if you wanted to put the true subject or actor at the end of the sentence, or if you wanted to avoid a dangling construction. In general, though, seven out of ten of your first-draft passives should be revised.

In still other passive constructions, the true subject or actor is unaccountably concealed in a prepositional phrase beginning with *in, at, for, with,* or *throughout.* The most tricky case is that in which the prepositional phrase introduces the sentence.

IN

NO:	*In the next chapter* the error messages *are explained.*
YES:	*The next chapter explains* the error messages.

AT

NO:	*At six of the campuses* facilities management contractors *are used.*
YES:	*Six of the campuses use* facilities management contractors.

FOR

NO:	*For managers* without formal training in accounting, a special training module *is needed.*
YES:	*Managers* without formal training in accounting *need* a special training module.

In some passive sentences, the ones I call "the invisible man," there is no mention of the true actor or subject. Sometimes such sentences are used to make statements in which it doesn't matter who the true subject is. (This last sentence is an example.) In other cases, though, writers unwilling to assert themselves, or to accept responsibility for a claim, make themselves invisible.

NO:	Weaknesses were unexpectedly discovered during the third phase of the study.
	WHO DISCOVERED THE WEAKNESSES?
NO:	The transition is intended to produce no disruption of service.
	WHO IS MAKING THIS PROMISE?
NO:	Your application has been rejected.
	WHO REJECTED IT?

Use "the invisible man" only when the subject does not matter (as in abstracts of papers or certain descriptions of procedures). Otherwise, name the subject.

One of the most characteristic signs of badly edited technical writing is the sentence that begins with an "it" passive. These introductory clauses can usually be scratched out without any loss of meaning. Otherwise, replace them with active constructions. Avoid

- It has been determined that . . .
- It can be seen that . . .
- It was found that . . .
- It was generally held that . . .
- It is believed that . . .

NO:	*It has been determined that we will* bid on the RFP.
YES:	*We decided to* bid on the RFP.
	Or
YES:	*We will* bid on the RFP.

"It" passives make your writing seem not only impersonal but inhuman. Look at this sample:

BEFORE:	*It was expressed* strongly that making the system as simple and easy to use as possible is an integral part of the CALC development. *It was accepted* that doing this was not a requirement for the initial prototype but that some proof would be necessary before further development could take place.
AFTER:	CALC, they insisted, must be as easy to use as possible. And they warned that unless we could prove the system was going to be simple, they would stop us from going beyond the prototype.

The rule against using *I* in formal scientific papers does not extend to ordinary business writing. Use pronouns and write in the active voice. And use verbs that show feeling and intention.

Put Words, Phrases, and Clauses in Parallel

The words *and, or,* and *nor* cause the reader to expect that what will follow the *and, or,* or *nor* is parallel in grammatical form to what preceded. One or two adjectives before an *and* call for an adjective after the *and.* One or two nouns before an *or* call for a noun after the *or.*

	PARALLEL ADJECTIVES
NO:	This new printer is faster, quieter, and will print graphs with higher resolution.
YES:	This new printer is faster, quieter, and higher in resolution.
	PARALLEL NOUNS
NO:	The president thanked them for their loyalty, dedication, and because they worked through the power crisis.
YES:	The president thanked them for their loyalty, dedication, and willingness to work through the power crisis.

A good writer will actually remove information from a sentence or passage to make it parallel, easier to read. Sentences that are out-of-parallel almost always force the reader to restart the sentence or clause in which the problem
occurs.

Items in lists—with numbers or letters or "bullets" to set them off—should always be in parallel, even if some information needs to be stripped away to make them so. In a list of four or five items, usually two or three will be in parallel form in the first draft; the quickest way to edit is to generalize the form of the few parallel items to all the others in the list.

NO: We'll know the new productivity program is working when we see the staff

- arriving at work on time
- correct their own mistakes
- stay out of accidents
- less absenteeism

YES:

- arriving at work on time
- correcting their own mistakes
- staying out of accidents
- using fewer sick leaves

Or

YES: . . . when we see

- less lateness
- less absenteeism
- fewer accidents
- fewer mistakes

Again, the more information you strip away, the easier it is to follow the list. The editorial choice is yours.

Emphasize and Contrast Your Points

Many of my clients tell me that they have trouble "making a point." Often, their problem is that they neglect to add small but essential words of emphasis. Notice how much more cogent each of these sentences becomes when one small word is added.

BEFORE:	The government admits that this project is a waste of money.
AFTER:	*Even* the government admits that this project is a waste of money.

BEFORE:	They propose to run the data through a contingency table analysis.
AFTER:	They propose to run the data through *just* a contingency table analysis.
BEFORE:	You can have ease of file creation or ease of report generation.
AFTER:	You can have *either* ease of file creation or ease of report generation. (But not both.)

Emphasis words are extremely useful, but a little tricky. Many writers put them in the wrong position—a subject that comes up again in Chapter 13.

Often, the best way to emphasize something is to both say what it *is* and say what it *is not.* To explain a feature, an action, or a method, describe its rejected alternative. Use contrast.

BEFORE:	The standard report shows only quarterly and yearly summaries. (Why "only"?)
AFTER:	The standard report shows only quarterly and yearly summaries, *but not monthly summaries.*
BEFORE:	These models can be used even by nonfinancial executives. (Why "even"?)
AFTER:	These models can be used *not only by financial executives* but also by nonfinancial executives.
BEFORE:	We want better-trained operators and shorter training programs. (As opposed to what?)
AFTER:	We want better-trained operators, but we want shorter, *not longer,* training programs.

Adding contrast may add "unnecessary words," but, again, they may be necessary to guide the reader's thoughts and make your point!

Do Not Be Too Compressed or Complicated

By now we have mentioned most of the first-draft flaws that can make your sentences 10 or 20 or even 30 percent longer than they need to be. Most people who complain that their sentences are too long (or who hear that complaint from others) really have a *wordiness* problem, not a sentence problem.

There are some writers, though, who just put too damned much into a sentence: too many clauses, too many data, too many qualifications and exceptions. These sentences might be fine for a machine to read, but they are unfit for human consumption. They should be broken into smaller, more manageable chunks. Here's a good example:

BEFORE:	In addition to solid, dashed, phantom, centerline, and invisible line fonts, numerous linestring fonts are available that provide generation about a centerline with variable spacing (width), layer of insertion options, and left, right, and center justifications.
AFTER:	The available line fonts are solid, dashed, phantom, centerline, and invisible. There are also *linestring* fonts that generate about a centerline. These linestring fonts can vary spacing (width), insert layers of information, and justify text to the right, left, or center.

And here is one I still cannot decipher:

> Identical assumptions were made with respect to the value of critical variables for each of the alternatives in order to provide a basis for comparison of cost differences due strictly to factors related to differences among the alternative configurations.

Sometimes even a very small bit of extra information can make a sentence harder to understand than it needs to be. Consider this sentence:

> The only way to speed monthly, quarterly, and year-end closings (other than for inventory accounts) is to move up all significant cutoffs.

The contingency—*(other than for inventory accounts)*—is an exception that can be stated in the *next sentence,* leaving us with:

> The only way to speed monthly, quarterly, and year-end closings is to move up all significant cutoffs.

Most of the wordiness in sentences, as I have said, is due less to the structure of the sentences than to the words and phrases used. But many sentences—even short ones—are more complicated than they need to be. Consider these examples from a software user's manual:

COMPLICATED

- To find what file names have been assigned, DFIL should be entered.
- If the user wishes to find what file names have been assigned, he or she should enter DFIL.
- If one wishes to find what file names have been assigned, one should enter DFIL.
- The command DFIL can be used to find what file names have been assigned.
- Users who wish to find what file names have been assigned should enter DFIL.

- The listing of assigned file names is provided by the command DFIL.

PAINLESSLY SIMPLE

- To find what files names have been assigned, enter DFIL.

Do Not Be Too Childlike and Simplistic

Sometimes, after I have spent a day or two exhorting my clients to write with plain, simple sentences, free of irrelevant detail and showoff language, I learn that some of my clients do not suffer from these afflictions at all. Quite the contrary, their problem is *too much simplicity:* short, one-clause sentences, primer language, and a tedious repetitive style that would bore any reasonably bright reader.

These laconic writers (who, by the way, are usually convinced that what they really need is more big words to make them sound more "flowery") find it hard to distinguish between clarity on the one hand and childish oversimplicity on the other. They do not know that unusually short sentences (fewer than ten words) are a novelty in professional writing, used for variety or to make an emphatic point.

Most sentences that are less than eight or ten words long are too short and simple for adult readers. Unless they contain powerful information, they seem abrupt, choppy, childlike. For example:

BEFORE: The process is very simple. Each file is described in terms of its attributes. Each attribute can be changed. There is a design command for each attribute. To change a file, change one or more attributes.

AFTER: Put simply, each file is defined as a set of attributes, any of which can be changed. Because there is a command uniquely associated with each attribute, you redefine the file by entering the appropriate commands and changing the corresponding attributes.

For most readers, strings of short sentences seem choppy and distracting. They are acceptable, however, for readers with poor reading skills or limited education.

An unrelenting use of simple sentences, all starting with the subject, is clear but boring. Consider this case:

BEFORE: The planning system has three components. The first component is an assessment procedure. The second component is a forecasting procedure. (This component uses straight-line trends.) The third component is a decision process. This process is of the heuristic type. It does not use optimization techniques. Linear programming was tried in an earlier version. However, the users decided that LP had too many assumptions and was not realistic.

AFTER:	The planning system has three components: assessment, forecasting (using straight-line trends), and decision. The decision component uses a heuristic procedure rather than optimum-seeking methods—rejected when an early group of users decided that linear programming called for too many unrealistic assumptions.

Note that the Before is *easier* to read. But it is clumsy and tedious. Intelligent readers will find the first paragraph slow and boring, especially if they already know something about this subject. Probably the "best" style, depending on the sophistication of your reader, is somewhere between the two versions.

Be careful. When writers given to this excessive simplicity try to compose longer, more-complex sentences, they tend to join these primer clauses with the weakest of all links: *and.*

And is the weakest and most dangerous of all linking words. When two clauses are connected with an *and,* we know nothing more than that they are related somehow. (*And* means little more than a semicolon.)

Look for more-interesting connections. In particular, look for complex sentences.

NO:	The system uses twenty commands *and* most users can learn these in two hours. (What does this mean?)
YES:	1. Because the system uses only twenty commands, most users can learn these in just two hours. 2. Even though the system uses twenty commands, most users can learn them in just two hours. 3. Not only does the system use twenty commands but most users can also learn them in just two hours. 4. The system uses just twenty commands, but, typically, users can learn them in only two hours. 5. Comprising only twenty commands, the system can be learned in about two hours. 6. Despite its twenty separate commands, the system can be learned in only two hours. (Or does it mean something else?)

Vary Your Sentence Patterns

Quick. Tell me which sentence patterns you use.

You cannot, I am sure.

Most of us have been writing in the same few basic sentence patterns for so long that we are unaware of what they are. Composing sentences is mainly a habit rather than a deliberate craft.

Yet, good writers use more than just a few basic sentence forms. Their sentences vary in length and structure. There are occasional surprises, interesting turns, clever ways of putting things.

If you can write one good sentence, your next editorial concern should be for the variety of your sentence forms and patterns. You must care not only for the merits of your separate, individual sentences but for the overall effect.

Consider this pair of paragraphs:

A. Too many of our sentences are in the subject-verb-object or subject-verb-complement pattern. Most people use this pattern in nine out of ten of their sentences. This practice makes our writing dull. The practice also limits our ability to express complex relationships. This kind of writing limits our ability to use emphasis. We cannot build suspense and we cannot guide the reader through the subtleties of our thought. This recurrent pattern bores most readers. It is boring you right now.

B. In contrast, more effective writers—knowing how sentence patterns can be used to guide and inform the reader—let no more than 50 or 60 percent of their sentences fall into the S-V-O or S-V-C pattern. Ever mindful of their readers, and knowing the dangers of monotony, these writers have learned the key to an interesting style: variety.

By almost any standards, paragraph A is less interesting than paragraph B. True, paragraph B is *harder to read;* the syntax forces the reader to wait longer between full pauses or stops. But paragraph B is *harder by design;* its sentences are crafted, tooled by an editor. They are not like the tangled, foggy, wordy messes I see in many first drafts—and even in some published ones.

If you want to be a better–than–average writer, you will have to vary your sentences more than the average writer does. You may even have to do things that distress the less-imaginative technical editor. For example:

STARTING WITH A PREPOSITIONAL PHRASE

BEFORE:	This contract cannot be extended without a new project director.
AFTER:	Without a new project director, this contract cannot be extended.

STARTING WITH APPOSITIVES

BEFORE:	Clearer reports, sharper graphics, and faster turnaround are the expected advantages of the new project management system.
AFTER:	Clearer reports, sharper graphics, faster turnaround—these are the expected advantages of the new project management system.

STARTING WITH A SERIES OF DEPENDENT CLAUSES, WITH DELIBERATE RHETORICAL REPETITION

BEFORE:	We have submitted four questionnaires to OMB. They took at least four months to approve each one. This is the most-

	complicated instrument we have ever proposed. Consequently, we expect OMB to take longer than four months with the latest one.
AFTER:	Because we have already submitted questionnaires to OMB, and because OMB took four months to approve each of them, and because this questionnaire is the most complicated of them all—because of all this, we are certain that OMB will take longer than four months to approve our latest questionnaire.

Of course, this last example is an odd case. By converting from the basic, pedestrian format in the Before version, we give the sentence a new tone, a new effect. The sentence becomes longer—though easy to read—and the deliberate repetition even suggests the wearying process of getting OMB (the federal government's agency for approving questionnaires) to approve an instrument.

There are dozens of other patterns you can create. Instead of putting the dependent clauses at either the beginning or the end, provided you do not do it too often, you can put them in the *middle.*

BEFORE:	This plan will never satisfy the sponsor, even though it is popular in the engineering division. (backward; reverse the sequence)
BETTER:	Even though this plan is popular in the engineering division, it will never satisfy the sponsor.
BEST:	This plan, even though it is popular in the engineering division, will never satisfy the sponsor.

How many of these odd or nonstandard sentences you use is up to you. More specifically, you must choose whether as few as 10 percent or as many as 50 percent of your sentences will be something other than the conventional, simple S-V-O or S-V-C pattern. You will have to decide, for example, whether you want to change two simple sentences into one complex sentence.

BEFORE:	The guidelines were published late. As a result, we have not yet selected the value-engineering consultant.
AFTER:	Because the guidelines were published so late, we have not yet selected the value-engineering consultant.
BEFORE:	The guidelines were not available to us. However, we were able to select a consultant in time.
AFTER:	Even though the guidelines were not available to us, we were still able to select a consultant in time.

Do not misunderstand me. After telling you for many pages to make your sentences simpler and clearer, I am not suddenly changing my position. On the contrary. One of the advantages of learning the editing techniques in the preceding two chapters is that you can often trim a long-winded sentence into a sentence that is so short that it is too short for your reader. When that happens, you can often turn two pared-down, primer sentences into one interesting adult sentence. Consider this case:

BEFORE:	At that point in time the equipment which was available to us could not be utilized to perform investigations of the presence of AOK in densities which were less than one part per thousand. Pursuant to this limitation we reached a decision that we would not be in a position to implement our enforcement of the standard which had been proposed.
INTERMEDIATE:	The equipment available then could not detect AOK in densities less than one part per thousand. So we decided that we could not enforce the proposed standard.
RESULT:	Because the equipment available then could not detect AOK in densities less than one part per thousand, we decided that we could not enforce the proposed standard.

The "final" version is made possible by the intermediate editing of the first two sentences. The result is a long sentence (twenty-seven words), but one so easy to follow that most readers intelligent enough to be reading about this subject would be able to follow this pattern. (If, in contrast, you had tried to combine the first two into a single complex sentence, the result would have been monstrous and unreadable. For fun, try it.)

Again, the aim is variety. Once you believe that you can write a long sentence well, then allow yourself an occasional long sentence. Save your very short sentences for high-impact, dramatic materials. Use them occasionally to break up the boring repetition of fifteen- to twenty-five-word sentences.

And fragments! Acceptable once in a while—if you write them on purpose.

Decide whether you want your writing to "sound" different from the avowedly tedious and predictable style of most technical, business, and government writing. If you do, then you will have to put more variety in your sentences, even insist on some devices that technical editors dislike (such as beginning about 40 percent of your sentences with something other than the subject). Short of writing in a way that sounds odd or difficult to your reader (which would conflict with your purpose for writing and editing), the choice is yours.

If you are content to be among the competent, average writers, that is up to you. But if you want to be among the excellent writers—the writers that readers pay most attention to—you must learn to vary the length and form of your sentences.

HOW TO EDIT A PARAGRAPH

A paragraph is a group of sentences—usually between three and six sentences—organized around a focal sentence. The key to writing successful paragraphs is, first, to make sure that the focal sentence is clear and distinct; and, second, to make sure that every other sentence in the paragraph is tied to it in a logical, evident way.

The focal sentence in a paragraph, often called the "topic sentence" or "thesis sentence," is usually the first sentence in the paragraph. Usually, but not always. A paragraph may begin with a preliminary or transitional sentence or two, followed by the focal sentence. Or, indeed, a paragraph may build slowly and gradually through five or six sentences until it reaches a thesis sentence at the end.

Whether the focal sentence comes first or last, you must still write your paragraph in a way that makes it more and more interesting as it develops. If a paragraph begins with a focal sentence, then the rest of the paragraph contains increasingly more interesting discussions or implications of the focal theme. Of course, if the paragraph ends with the focal sentence, then the paragraph builds toward that climactic idea.

In either case, the last part of a paragraph is the best-remembered part, and it is the part that motivates the next paragraph. If you are in the habit of writing good first sentences and nothing more, your writing is probably dull and ineffective. (Unless, of course, you are only writing for skimmers.)

Typically the focus or topic or theme of a paragraph is at the beginning. And the pattern of the paragraph emanates and develops from that point. Paragraphs follow a few basic patterns, depending upon the relationship between the focal sentence and the rest of the paragraph. Most paragraphs do one of the following:

- Explain or illustrate the focal sentence
- Give examples or applications of a general statement
- State contingencies, qualifications, and exceptions to the focal claim
- Compare or contrast the focal idea with some other idea
- Teach the reader to use or apply the focal idea
- Link the focal idea to the focal idea of some earlier or later paragraph
- Define key terms in the focal sentence
- Defend or rationalize the thesis or argument of the paragraph
- Detail the implications of the focal sentence

If you find a first-draft paragraph that does not fit any of these patterns of development, then you should be able to name or describe *your* pattern. If you cannot say what the paragraph *does,* how all the other sentences are connected to the focal sentence, then your paragraph needs work.

Organizing and Connecting Sentences

Most of the improvements you will make in your paragraphs will be small additions: deliberately repeated words, transitional words (like *for example*) that bind one sentence to the next. Consider these examples of pointed repetition of key words.

WEAK:	Jones objected that the new carrier would charge twice as much and give no better service. But it included insurance premiums previously billed separately.
STRONGER:	Jones objected that the new carrier would charge twice as much and give no better service. But his *objection* overlooked the fact that the *charge* includes insurance premiums . . .
WEAK:	Heuristic decision models solve problems for which there are no optimization techniques. These are described in the next chapter.
STRONGER:	Heuristic decision models solve problems for which there are no optimization techniques. *These heuristic models* are described . . .
WEAK:	When you try to sell an MIS to some managers, you learn that they do not use quantitative management techniques. This makes a problem for the salesperson.
STRONGER:	When you try to sell an MIS to some managers, you learn that *these managers* do not use quantitative management techniques. This *lack of interest* in data makes a problem for the salespeople.

Pointer words—*this, that, these, those,* and the like—are essential for the linking of ideas. Often, though, they are not enough. Be sure that these pointers point unmistakably at one noun or phrase. When in doubt, repeat the noun or find some other construction.

Usually, the easiest way to connect two sentences in a paragraph is with transitional words or phrases. In most well-edited paragraphs, as many as 20 or 25 percent of the sentences begin with such a phrase.

The most useful English transitionals follow.

TO EXTEND OR ELABORATE:

then
first, second,
finally
in fact
for example
further
moreover
next
soon
in addition, additionally

TO ASSERT CONSEQUENCE:

accordingly
consequently
thus, therefore
in short, in sum
in conclusion, to conclude
as a result
for these reasons
given
then

TO REFER:

from this
meanwhile
after that
before that
lately, recently
formerly, previously

TO CONCLUDE:

of course
even so
granted
admittedly
still
after all

TO UNDERSCORE SIMILARITIES:

similarly
likewise
in the same way
just as
at the same time
along these lines
by analogy
in a comparable way, comparably

TO HIGHLIGHT:

to be sure
indeed
certainly
clearly

TO CONTRAST:

yet, still
however, but
nevertheless, nonetheless
on the contrary, in contrast
on the other hand
although, though
albeit
notwithstanding

Watch how much clearer paragraphs can become with the addition of just a few small words.

BEFORE: I believe in management by objectives. My boss often remembers my objectives but forgets to give the budget to carry them out. I've concluded that it won't work unless authority is tied to responsibility.

AFTER: I believe in management by objectives. *Unfortunately, though,* my boss often remembers my objectives but forgets to give *me* the budget to carry them out. *Consequently,* I've decided that *MBO* won't work unless authority is tied to responsibility.

BEFORE: Metering is recognized by the water industry as the most effective means of eliminating waste because it communicates the exact amount and cost of water directly to the customer. The demand for water in the city of Boulder, Colorado, dropped 35 percent when they converted to metered billing. The effect is found in California where the ten-year average for the metered Los Angeles water system at 176 gallons per capita per day is over 100 gallons less than the consumption for the unmetered city of Sacramento.

AFTER: Metering is recognized by the water industry as the most effective means of reducing waste because it communicates the exact amount and cost of water directly to the customer. *To illustrate,* the demand for water in the city of Boulder, Colorado, dropped 35 percent after it converted to metered billing. *Similarly,* in California, the *demand for water* in the metered Los Angeles system has averaged at least 100 gallons (per person, per day) less than in Sacramento, where the system is unmetered.

BEFORE: The department's long-range Leak Program has resulted in system leakage considerably below the national average. Our 6.4 percent unaccounted water is well below the 12 percent national figure. Leakage is one part of the 6.4 percent figure which includes meter discrepancies, reservoir seepage, firefighting uses, and other losses. However, additional efforts are being made to reduce leaks. Technicians have already received training in the use of new electronic surveillance devices and are now in the field full time to reduce leaks in the system.

AFTER: The department's long-range Leak Program has reduced our water leakage to a level considerably below the national average. *Specifically,* our 6.4 percent rate of unaccounted water is well below the 12 percent national figure. (*Of course,* "unaccounted water" includes factors other than leakage.) *In addition,* we are now training our technicians in the use of new electronic surveillance devices that are *already* being used to reduce leaks in the system.

BEFORE:	Most people believe that unregulated trade cannot work equitably in any but the smallest community. It has never been tried. It might work. The fact that something has never been tried is not a defense that it will work. That would be absurd.
AFTER:	*Although* unregulated trade has never been tried on a large scale, most people believe that it cannot work equitably in any but the smallest community. *Of course,* it would be absurd to defend an idea strictly on the grounds that it has never been tried. *Nevertheless,* I believe *unregulated trade might work!*
BEFORE:	Psychologists say that if we are not true to ourselves we become ill. They say that everyone acts according to his or her needs all the time. They say the purpose of counseling is to get in touch with yourself. They say you must do what you want to do. When you want to stop seeing the psychologist, they say you are running away.
AFTER:	*On the one hand,* psychologists say that if we are not true to ourselves we become ill. *On the other hand, the same psychologists* say that everyone is true to his or her needs all of the time anyway. They say that the purpose of their counseling is to put you in touch with yourself *so that* you can do what you want to do. *If what you want to do, however, is stop seeing them,* they say you are running away.

The most popular of the transitional words is *however.* The word *however,* like several of the transitionals, is more effective a few words into the sentence than at the very beginning. (The terms *in contrast, on the other hand,* and *though* work similarly.)

Put *however* after the word or phrase that you are contrasting with the previous sentence.

GIVEN:	Most writers put *however* at the beginning of a sentence.
NO:	However, I prefer it later in the sentence.
YES:	I, however, prefer it later in the sentence.
GIVEN:	Most data-base management systems have complicated rules for the structure of files.
NO:	However, INFO-III has flexible rules for file structure.
YES:	INFO-III, though, has flexible rules . . .

Sometimes you will have to recast the second sentence a bit so that the *however* or other word can appear within the first few words. Strictly speaking, however, you can put *however* anywhere. It seems strange when it is the *last* word, however.

You may have noticed that throughout this book I have never used *however* (the transitional form) at the beginning of a sentence. But I have often used *but* in the place that many writers would put the *however.* Now I know that your elementary-school teachers expressly forbade you to put *but* at the beginning of a sentence, but, even so, many contemporary writers are using *but* as a lighter, quicker substitute for *however.* And they are using *and* as a replacement for *in addition* or *moreover.*

Use transitionals intelligently and boldly. As long as you do not overuse them or use them in sentences where they are unnecessary, you need not be intimidated by the editor or colleague who finds them "flowery." They are not flowery; they are basic, effective words that clarify the logic of your paragraphs. Fight for them.

Length

Paragraphs can literally be as long as you want them to be. A good writer will vary the lengths. Indeed, a page on which all the paragraphs are of the same length looks as though it might have been written by a machine rather than by a person.

Until recently, one-sentence paragraphs were suspect. Except for dramatic announcements ("The project could not be saved!") or functional information ("The tables appear in Chapter 4"), the one-sentence paragraph was of little use.

Modern editors, though, like shorter and shorter paragraphs. Tabloid newspapers, and other periodicals written at the sixth- or seventh-grade reading level, are filled with one-sentence paragraphs. In fact, few of their paragraphs have more than two sentences.

The reason for this trend is clear. Everyone hates long paragraphs. Every man, woman, and child on the planet. Because readers usually will not lift their eyes and take a deep breath until the end of a paragraph, long paragraphs fill the reader with fears of suffocation.

So, my advice must be qualified. Vary the lengths of your paragraphs, subject to an upper limit or length. Letters should have few paragraphs longer than eight typed lines; memos, eight or ten typed lines. Formal reports can stand a ten- to twelve-line paragraph; scientific papers and scholarly treatises, up to fifteen. Beyond these limits, you tax the abilities of even your most patient reader.

Notice that I talk about paragraph lengths in terms of numbers of typed lines. My hunch is that this is the most valid way to gauge a paragraph. As a result, since smaller type gives you more words to the line, smaller type allows you to produce longer paragraphs without bad effects. (There is a slight offsetting disadvantage for small type because it is somewhat harder to see.)

In any event, never allow your paragraphs to become immense. The following specimen is not the worst I've seen, but it is bad enough.

BEFORE: HSAs and SHPDAs have a continuing responsibility to assemble relevant data both for the development of plans and to assist in the performance of their respective review activities. Heavy emphasis is placed in the federal guidelines, for example, on the identification and quantification of "indicators" of health status and health systems status as indices upon which roles for achievement are to be established. On the other hand, both Public Law 93-641 and the federal regulations have made clear that the maintenance and collection of data should be based to the maximum extent possible on existing sources. The Center for Health Planning can contribute significantly to the organization by both state and regional agencies of efficient systems for the compilation, management, and access to relevant data necessary in the performance of those agency functions. It can, in addition to describing appropriate types and sources of data, advise as to appropriate methods of storing and filing to permit optimum accessibility and utility. The Center's principal role in this connection, however, would be providing insights to the planning agencies into the relative significance of different types of data available, analytical techniques which would be appropriately applied to such data, methods of interpretation adaptable to fundamental tasks of planning and review which the agencies must carry out. Thus, it is vital that the Center have significant expertise and familiarity with not only the techniques of data management and analysis but more particularly in management and analysis of data pertaining to health status and health systems planning and plan implementation. This is particularly critical in terms of one of the primary responsibilities of the Center, to communicate to the volunteer participants in the planning process the relevant skills so important to their gaining an understanding of the issues and techniques vital to sound health planning.

Now, I know you did not read that paragraph. No one would read it without being forced to.

And that is my point. Hidden in this intimidating mess of characters is an important selling theme, a key part of a proposal to the Public Health Service.

If you examine the paragraph (and I really do not expect you to), you will see four subtopics:

1. HSAs and SHPDAs must gather considerable data to fulfill their legal responsibilities.
2. They are restricted, for the most part, to other people's data (existing sources).
3. They need help in
   - Getting and managing the data
   - Using and interpreting the data

4. Therefore, the people who help them (the center) must be skilled in both data management in general and agency functions in particular.

Studying this list of topics, you see one of the really unfortunate ironies of the long paragraph: namely, writers who are well into a long paragraph rush through the last topics in an effort to finish. As a result, long paragraphs often compel writers to write *too little* about matters that need more explanation. Here is the revised version of the paragraph:

AFTER:

HSAs and SHPDAs are required by federal law and guidelines to assemble data for planning and review. For example, they must develop health status and health system indicators—scales that can be used to set goals and to describe progress toward the goals.

Despite the heavy data burden placed on these agencies, the federal guidelines also restrict the agencies to existing data, data gathered by other agencies, usually for unrelated purposes. As a result, the agencies will need technical assistance in securing the available data and applying it to their own needs.

The Center for Health Planning should provide that help. It should devise systems for

- Compiling and organizing relevant data, and
- Filing and storing data so as to ensure access.

In addition, the center must advise and train the volunteer boards of the planning agencies in the best techniques of analysis and interpretation.

To do this job well, the center must have on its staff people who are expert both in the general techniques of data acquisition and management and in the particular applications of those data to the functions of the HSA and SHPDA.

In the rare event that your very long paragraph does not lend itself to any kind of logical subdivision into smaller paragraphs, then—I tremble to say this—break it anywhere. Even though your paragraph is an intact whole, and even though any reasonable, careful reader will see it, nevertheless the prejudice against long paragraphs is so powerful that it is almost a physical barrier to communication. Long paragraphs are either not read or not read carefully by the vast majority of readers. Stop writing them.

12

Exhibits and Artwork

THE GREAT DEBATE

Several times in these chapters about editing I have used the verb *presume.* I have urged you to presume against long words, to presume against foreign terms and clichés, to presume against long paragraphs. Later I shall tell you to presume *in favor of* optional commas.

When I urge you to presume for or against something, I am not commanding you to write in a certain way. Rather, I am proposing the little arguments you should have with yourself. When you presume *against* something, you try to talk yourself out of it. (If you can't, it stays.) When you presume *in favor of* something, you try to talk yourself into it. (If you can't, it goes.)

Now, I don't know exactly how to advise you on the general question of exhibits and artwork. Obviously, those of you who never use a chart or figure should probably use more—and vice versa. But on the question of whether diagrams and tables are automatically a good thing, whether a picture is worth ten thousand words, or one thousand words, or even ten words—on this greater question I am stumped.

In my experience, most writers like to use pictures and tables; some even compose their first drafts by drawing the sketches and composing the text to go with them. And most readers seem to appreciate pictures and tables, often turning to them first as a kind of summary or digest of the longer report. Despite this preference, many of these pictures are cryptic, and even more are gratuitous, not really helpful in clarifying the writer's point.

The issue is complicated. On the one hand, exhibits (by which I mean anything other than ordinary text) are useful if they do nothing more than break up the visual monotony of the paragraphs. And several times I have urged that you must—for the sake of overcoming the boredom and fatigue of the reader—put some graphic interest into your writing. On the other hand, many of these graphic embellishments are just ornaments, and a few, filled with dense equations that could have been better expressed in plain English, intimidate a reader who would otherwise have understood without difficulty.

Then there are matters of taste. I know a gifted management scientist who loves articles that make their points with cartoons—even lame, amateur, stick-figure cartoons! For me, though, the thing I dislike most is attempts at humor by people who are less than exceedingly funny.

The debate about exhibits is also a sometimes fierce dispute over costs and effort. Between the writer and the reader there are about a half-dozen people vehemently opposed to exhibits: the typists, the publications manager, the editors, and even the artists (unless they are paid piecework). If you are writing for publication, you can expect that the editors of most technical journals and the publishers of most professional and technical books will oppose your exhibits as well.

Obviously, exhibits—charts, tables, figures, photos—cost more money and demand more effort than "straight typing" or "straight typesetting." A typist who can do ten crisp, correct pages in two hours will need the same time to produce a one-page table. (Even the new electronic office typewriters, which can align and justify tables, still take time to learn and use properly.) A typesetter will take as long setting six pages of journal text as setting three inches of equations.

In effect, the production process resists exhibits far more than it resists text. The various frictions that slow the progress of your report or proposal are greater, for example, when you offer a network depiction of your schedule than when you offer a simple list of the start and finish dates of your projected activities. Indeed, I know of typists who object even to "bullets," the big dots used to set off items in a list, because making the bullets forces the typist to go through the extra step of filling in the lowercase *o*'s with a pen.

Because there is so much resistance to exhibits, and because—when time is running out—the abandoning of exhibits is the most attractive timesaver, and because exhibits can cost more money than anyone wants to spend . . . for these reasons you will probably have to plan for more exhibits than you expect to get. Like a bargainer, you will have to think of some of your figures and diagrams as bargaining chips, items you are willing to give (or trade) away when the occasion arises.

Do *not* begin your planning and drafting, therefore, with a conservative estimate of the number of exhibits you will need and of the time and budget needed to produce them. Most of us have no idea of how long it takes to draw symmetrical arrows or paste up a title page in which all the lines of type are parallel. If you begin with too little in your "exhibits budget," you may be forced to give up some graphics that matter to you.

But how can you know which ones matter? How can you decide where an exhibit is necessary or merely desirable? When can you feel secure in your dispute with an editor who insists that a certain table is a waste of space?

USE AN EXHIBIT WHEN THERE IS SOMETHING TO SEE, RATHER THAN JUST SOMETHING TO READ. When you want the reader to notice a pattern, a relationship, a function, a tendency, a comparison, or any other configuration that cannot be rendered in a sentence or two, create a table or chart or figure. When you cannot describe an object or process in a way that permits the reader to see the connection between the parts of the thing and the whole thing, use a picture or photo or flow chart.

Exhibits give *synoptic* information. They show the reader parts and wholes, causes and effects, systems and environments. Sentences *can* do all these things, but if it takes a lot of words to explain or describe what can be visualized in an instant, then use the visual information as well.

Unfortunately, this criterion is neither precise enough nor inclusive enough to answer all your questions about where graphics are needed. Suppose a graph is just a "slightly better" way to represent a relationship. Or suppose your reader "likes pictures." Or suppose your text is visually boring and fairly cries out for something other than paragraphs. What do you do then?

(Or suppose you have a salaried artist with not enough work to keep him or her busy. Will you let that resource go unexploited?)

Ultimately, the battle over exhibits is between you and your publications/production people. No matter how it comes out, though, keep these principles in mind.

KEYS FOR MORE-EFFECTIVE EXHIBITS

Whichever exhibits you use—tables, flow charts, maps, schematics, snapshots, or CRT displays—there are a few rules that most editors and publishers abide by.

Get the highest production quality you can afford. In much the same way that careless writing calls attention to the writer instead of the message, careless exhibits call attention to the artlessness of the production instead of the content of the graph or figure. At any budget, and with any resources, you can produce better exhibits than you think. (Later in this chapter I'll suggest ways to make amateur art look a bit more professional.) And if you are in a business that demands high-quality art, you *must* secure a level of talent and equipment that is at least as good as that of your competitors.

Even if your firm cannot afford elaborate in-house facilities for art and photography, you can find consultants and contractors to supply the work for you. America is filled with underemployed commercial and technical artists; big cities have scores of small art firms eager to work for you.

If your firm's product is *paper,* that is, if the main result of your work is plans, studies, or reports, be sure to secure professional art services (along with high-quality typing and copying equipment). A consulting engineer who produces bad-looking reports—regardless of their technical quality—will not long prevail.

Name and number every exhibit. Exhibits will usually pass through more hands than any other part of your report or manual. Be sure to name and number each exhibit, and refer to it by both name and number. And when you are marking the place in your text where the exhibit will eventually appear, refer to it by name and number again.

Remember. Exhibits are hard to manage. They move around—often going through review processes that differ from those for the rest of the text. They get shipped to other companies and are sent "under separate cover." Make sure they are always wearing their dogtags.

Similarly, make sure you, the writer, have a copy of each exhibit, even if it is only a rough copy. Often, while working on the text, you will need to refer to the table or diagram; be sure you can get it. And sometimes you will decide that some of the facts or terminology in the exhibit need to be changed. Only if you have your own copy will you know precisely what changes must be made. (Be prepared for howls from the art people when you tell them the changes you want. Artists tend to resist changes more than any other group in the organization —except for the programmers.)

Put exhibits in an accessible place. Any exhibit that appears in the body of the document (as opposed to an appendix or attachment) must be in the place where it is discussed. It should be as close as possible to the first mention of it, on the same page or next page if possible. If you are copying on both sides of the paper, then the preferred arrangement is to have the text and exhibit on facing pages (assuming you cannot put them on the very same page, the ideal arrangement) so that the reader can follow the text and see the exhibit at once.

The practice of separating text and exhibits, whatever its motivation, is completely incompatible with effective communication. If a reader is obliged to flip pages to see a picture of the process described in the text, the manual cannot be called well designed. That this convention is widespread, even commonplace, does not alter the fact: When exhibits are removed from the text, a severe burden is placed on the reader, who may abandon the document out of sheer frustration.

(A possible exception to the rule is the multivolume document in which all the figures are in one separate volume. In this case, the reader can open *two* documents and read them side by side. This is still awkward, but better than flipping back and forth.)

Write something about every exhibit. If you have nothing to say about a particular table or figure, something is wrong. Either you have failed to consider your reader (who probably cannot see what points you are making with the exhibit) or perhaps the exhibit does not belong in the body of the document at all! (Some scientists seem to feel that once they have assembled some data, at great cost, they are obliged to throw these data into the Results section—even if the data are irrelevant to the discussion.)

There is some debate about this last point. Because a well-made exhibit is self-sufficient (as I shall explain below), some prominent editors see nothing wrong with a Results section that contains nothing more than a table with the sentence "The results are shown in Table 4." Or a Schedule section that contains one figure with the sentence "Figure 5 shows the schedule of inspections." True, if the exhibit is well made, a thoughtful reader *could* learn everything he or she needs to know just from studying it. But will the reader study the exhibit? Not likely. Will the reader detect what you consider the main point, the key relationships, the most unusual aspects? Not likely.

If there is anything important to be inferred from the table or figure, say what it is. If there is nothing that needs to be said, then what is the exhibit doing there in the first place?

Make every exhibit self-sufficient. The reader should find everything needed to understand an exhibit within the exhibit itself. The title should be descriptive; the caption should be concise and readable. Any codes or abbreviations used in the exhibit (especially those used as column headings) should be explained in a legend or in footnotes; any symbol or graphic contrivance should be explained in a legend or key. (The exceptions, of course, are those abbreviations and symbols known universally among the audience of readers. But there are fewer of those universal conventions than you would think.)

Does this rule of self-sufficiency contradict the earlier rule, that every exhibit needs text to explain and discuss it? Not at all.

First, that fact that every aspect of an exhibit is understandable to the reader does not mean that the reader will notice what you want noticed or infer what you want inferred. In particular, the reader may not appreciate the place of the exhibit in the message you are developing. You will need text to point out, for example, that the circuit described in the diagram has fewer components than its predecessor. Or that the milestone dates on the network schedule are four weeks earlier than required in the plan. Or that the rate of change from column 2 to column 3 is only half the rate for column 3 to column 4.

Second, exhibits tend to be used and reproduced on their own. The flow chart you developed for a particular proposal may go into the company's "boilerplate" and be used in other documents. The table you prepared for one series of test results may be pulled out of context and inserted into some other report or compendium. Charts and pictures tend to outlive their original documents, partly because companies like to recoup some of the costs of preparing them. So, be sure that when your exhibit goes out on its own, you have given it all it needs to survive.

If possible, limit exhibits to one page. Because the main justification for using exhibits is to let the reader "see" something synoptically (everything at one glance), you defeat your purpose with figures and tables that take two or more pages.

Of course, there is a trade-off here. If we try to keep to one page, we often jam so much material into that small space that the reader needs a magnifying glass to understand it. (I have seen one-page figures in high-quality science magazines that were so dense with lines, symbols, and captions—hundreds of words of captions—that my first reaction was to laugh at them.)

Still, though, the better solution is to rethink the exhibit as two or three self-contained one-page exhibits, rather than as either a multipage table or an unreadably dense one-page figure.

(In this connection, beware of a small but dangerous trap. Because so many firms now have photocopy machines that can reduce the size of tables and figures, there is a tendency in some quarters to reduce past the point of legibility—especially when the words and numbers on the exhibit are reduced along with the artwork. Rather, if you are going to shrink your charts and pictures, type the words and numbers onto the *reduced version* so that the characters will be the same size as the text in the rest of the document.)

For large complicated charts, the *foldout* is the solution. Everyone in the production process hates foldouts, of course, but they are often the most effective way to present a complicated exhibit that cannot be "shrunk" to one page and should not be broken into two or three pages. Foldouts should be no wider than two ordinary pages, however; when they are bigger than that, they become as hard to refold as highway maps or collapsible umbrellas.

Most editors prefer, with good reason, that you present all exhibits on the same axes as the rest of the text, that the reader *not be forced to turn the report sideways to look at a picture.* If you decide to break this convention, though, be sure that the page is arranged so that the reader turns the book clockwise and the right-hand margin then becomes the *bottom* of the picture. Any other arrangement will oblige the reader to spin the book around *twice* to find his or her original place.

Do not violate the margins. If an exhibit appears on the same page with text, do not allow the exhibit to intrude into the margins of the page. If an exhibit is on a separate page, be sure the margins are no slimmer than those used for the rest of the document.

Although I know these rules may seem a nuisance—another irritating constraint on designing a readable chart or figure—they are for your own good. Anything that spills over into the margins looks horrible and gives your work a cheap, unplanned, cut-and-paste look. Your reader doubts the credibility of the person or company that produced it.

Also, there is the problem of binding. If your document is to be bound with plastic combs, or punched for a loose-leaf binder, then there is a risk that materials that go beyond the left-hand margin will be punched or obscured by the binding process. A manual in which one of your diagrams has a hole punched in it is a manual in trouble.

Put references and sources directly on the exhibit. When your table or figure comes from someone else's work, be sure the acknowledgment or citation is right on the exhibit itself, not elsewhere in the text. As I said earlier, your exhibits are often used by other people. If you have borrowed your exhibit from someone else, make sure the next borrower knows where the material came from.

Set your exhibits off. Exhibits should be highlighted with either dark borders or wide spaces. Because the reader deals differently with exhibits, he or she must be alerted to where the exhibits start and end. Borders (boxes) are the safest technique. Different typefaces, different colors—any device that makes the exhibit immediately distinct from the rest of the visual information on the page.

Be especially careful when there is a caption at the bottom of a figure. Be sure that there is a clear demarcation between the end of the caption and the beginning of the next text.

MAKING AMATEUR ART LOOK BETTER

If you work in a large, successful firm, you will probably have access to all the artists and all the graphics technology you need for your reports and manuals and proposals.

But suppose you work in a small, struggling firm. Suppose you work alone as an independent consultant. Suppose you are a college instructor with a kitchen-table business on the side. What then?

The best course is still to find an artist—perhaps another college instructor working from his or her own kitchen table. Even though not all artists are equally skillful, it is still true that artwork and exhibits prepared by professionals will nearly always look much better than our own amateur efforts. And if our amateur efforts look bad enough, we may fail to reach our professional objectives.

But even if you cannot or will not bring an artist in on your project, there are still ways you can make your amateur work look better.

First, equipment. Get a table suitable for artwork and graphics. The adjustable trestle table is best, but any wide, flat table will do. On that part of the table where you intend to draw, the surface must be hard, flat, and smooth—so that a line will not suddenly break or bump when you draw your pen into a hole.

DO NOT TRY TO DO GRAPHICS ON A DESK CLUTTERED WITH YOUR OTHER WORK. You need elbowroom to draw a flow chart. So, rather than clearing off your desk (a half-day project, no doubt), do your graphics work in another place. This will also spare you the embarrassment of spilling rubber cement on your appointment book.

Get yourself some basic drafting tools: a T-square, a 45° triangle, a 30°–60° triangle, a "French curve," a protractor, a high-quality compass.

Beyond these basic tools, acquire a set of *artist's templates,* plastic stencils that you can use to trace the shapes that appear most often in your drawings. At any good stationery or art supplies store you will find a rack of templates, covering every imaginable application from symbols in circuit diagrams to top views of shrubbery for site plans. You will also find templates with circles, ellipses, squares, and rectangles of every size, not to mention a dozen varieties of arrows. The more of these templates you own, the better; force yourself to buy one or two each month until your collection meets all your drawing needs.

You won't need to buy expensive artist's or draftsman's pens. (You may if you want to.) Today's business supply store has dozens of varieties of pens, and, as with the templates, you should keep acquiring a few each month until you have those few that feel best in your hand and are most responsive to what you want to draw. Note: Keep your art pens in a separate box from your other pens; don't make the mistake of wasting your favorite ballpoint on an unimportant memo.

The aspect of graphics in which you will have the most choices is *lettering.* Included among those plastic templates will be stencils for letters of several sizes and styles; you should have a few of them. Close by, on another counter at the office supply store, you will see mechanical lettering devices that permit you to work simply with a pen while varying the size, angle, and density of your letters through a few minor adjustments of the device. In some stores you may even find typemaking machines: inexpensive (that is, less than $1,000) versions of the much more expensive headline machines used by printers and typesetters.

Of course, the most popular approach to lettering these days is *transfer type,* letters that are rubbed from a sheet of letters and symbols onto your chart or table. With nothing more than a small wooden stick and a straightedge, you can produce high-quality titles and labels for your figures and charts—even headlines at the beginnings of your sections. The only problem with transfer lettering is that it is slow and subject to mistakes. Few amateurs can make a line of rubbed-off letters come out straight, and even fewer can get the letters evenly spaced.

Transfer letters are not the only transferable graphics. Scores of companies manufacture sheets of letters, symbols, borders, lines, curves, pictures—all of which can be rubbed onto your art. There are also hundreds of collections of "clip art" books filled with stock graphics and useful cartoons that can be cut and pasted into your pictures. (When you buy a clip art book, you also pay for permission to reproduce the pictures in it.)

And, joining the set of graphics that can be rubbed or clipped are hundreds of *tapes,* rolls of adhesive-backed lines, borders, arrows, symbols, which can be cut from the roll or peeled from a sheet and attached without so much as a pot of glue. These tapes (in more sizes and varieties than there are screws and bolts at the hardware store) produce such dark, clear, precisely rendered images that they have become more useful to the typical artist than a jar of India ink. With the right tapes, you could "draw" a critical path network with two hundred activities in it without even using a pen.

To complement your collection of templates, transfers, clips, and tapes, be sure you have a way of *removing* information as well. Acquire *white* tapes (in several widths) that can be slapped over your mistakes. Also, some "correction fluid," the chalky stuff used by typists to cover up errors. In this same category include some pencils and pens in that shade of blue that will be invisible to the photocopier, so that you can draw guide marks that will not reproduce.

Collect tablets of various kinds, with different line widths and different arrangements of columns. The more guidance you can get from the lines that are already ruled on the tablet, the less likely you are to prepare a table or chart that is misaligned.

In sum, if you cannot secure the services of a proper artist, then at least secure as many of the materials and devices a proper artist uses. Notice one critically important thing, though: *Using all these materials is slow and tedious.* Unless you take some pleasure in playing with graphics, the prospect of aligning and transferring four hundred or five hundred letters is oppressive, as is the prospect of cutting and fitting twenty yards of fine black tape. As you become impatient with it, you will become more careless and sloppy, so that eventually your exhibit will look worse than it would have with just freehand drawing.

The key to the problem, I think, is to find some person for whom this task would *not* be tedious and then bring him or her into the process. I am not suggesting that a high-school junior receiving minimum wages could do as good a job with your charts as an experienced technical artist; but the job would be better than if *you* had done it, and the young person so employed would certainly prefer this work to selling hamburgers in a fast-food restaurant.

Remember, my first recommendation is to find a professional artist. America is full of talented artists who are wasting their time on low-level cutting-and-pasting jobs, people who would be eager to add some quality to your written products. But if you cannot find one or afford one, then get someone for whom low-level cutting and pasting is a treat.

HOW TO TALK TO ARTISTS

The writer must tell the artist what to do. (By "artist" I mean anyone responsible for the final rendering of the exhibit, anyone from the typist to the photographer.) *You* set the specifications for the final product; *you* decide how it will work.

This is not to say that you would refuse advice from an artist, or fail to discuss your ideas with the photographer, or neglect to ask the typist for suggestions. Rather, remember that your graphics, like every other item of information in your message, are tied to *your* purpose and the needs of *your* audience. Even if you cannot visualize exactly how a chart should look, you should know what you want the reader to see in it and what you intend to write about it. Whatever the artist does, however, within *your* constraints, is up to him or her.

Before you express your wishes to the artist, know what you are talking about. Familiarize yourself with the equipment and resources in your firm (or in the artist's firm if it is a contractor). Know what your typewriters can do; know what graphics tricks your word processor can perform. Look at samples of work done by the artist.

The longer your experience with the artist, the better able you are to communicate your needs. Over time, for example, you learn how long it takes to complete certain tasks (and perhaps how much it costs). You learn that this particular artist (or typist or photographer) is especially good at certain things, but not at others.

You may also, in time, learn the correct names of the techniques and processes used in the art department—so that you cannot be "snowed" later by an artist who wants to get rid of you.

The subtlest things, the hardest things to detect, are the *values* of the artist, the aspects of the craft that matter most. You may learn, for example, that a particular illustrator believes "perspective" is necessary in certain kinds of drawings but disastrous in others. Or you may learn that a particular photographer will work tirelessly to find you what you want—until you happen to mention that a certain subject is lit improperly, at which point the photographer will do absolutely nothing other than exactly what he or she is told to do!

With this kind a background, you can communicate effectively with the artist. Describe:

- The *objective* of the communication (or that part of it the artist is working on) —say what informative, persuasive, or motivational end you are pursuing, and how that graphic fits in
- The *reader* or *audience*—tell the artist about those attitudes or interests that the reader will bring to the picture (including those beliefs and opinions you are trying to change)
- The *schedule* and *budget*—be sure the artist knows what he or she has to work with and when it *must* be done
- The *final disposition*—say how and where the piece will be used, as well as how it will be reproduced or printed

In short, help the artist to *realize your plan,* on time, under budget, and with the highest quality possible.

Editing IV—Checking for Grammar and Mechanics

COPY EDITING

By this stage in the process, you should have made all the important and substantive changes you are likely to make in your document. (Unless, of course, it has not yet gone around for review.)

Your editing is almost finished. In fact, the editing that is left can often be done by a competent secretary or typist. (Note, though, that many of the errors described below may have been *created* by typists.)

In the final stages, look for small mechanical problems:

- Words that have mistakenly been replaced with words that sound like them ("averse" for "adverse," for example)
- Small errors of grammar (like using "like" for "as")
- Incorrect or ineffective punctuation (like putting a period outside a quotation mark when you know that periods ALWAYS go inside quotation marks)
- Inconsistencies in your conventions (like court order, court-order, Court Order, Court-Order)

These small details, though less interesting than the grander questions of purpose and audience, though less challenging than the tough problems of logic and sentence structure, are nevertheless critical to your effectiveness. Clumsy errors of grammar ("Each of the problems *are* . . ."), or misspellings ("supercede"), or use of suspect words ("administrate") can cause your readers so much distress that they will reject your ideas and dread all your future messages.

No matter how dull the last minutes and hours of the process can be (and no matter how rushed you are), be vigilant. Get rid of those last few bugs.

Word Bugs

People who study language often assemble long volumes on "usage," collections of the terms that are most often misused or confused. Over a lifetime, a professional editor will learn a few dozen of the more problematical and typical cases (such as *as* versus *so*), perhaps even a few hundred. But only a small group of scholars knows them all.

The trouble with usage is that unless you know you have a problem, it never occurs to you to look up the problem in a reference. If you knew that nearly everyone misuses the word *hopefully,* you might check its meaning in a reliable source. If you knew that the word *anxious* is ambiguous, you might read about it and discover that *eager* would be a better choice.

But, alas, very few people who are not professional writers or editors see anything wrong with the expression *identical to*—and neither do you!

The only way to become aware of most usage problems is to read about them. This chapter includes some common problems and their solutions. But this array is elementary. To learn about usage you must read usage reference books *before the fact,* that is, before you need a ruling. In this way, you will become aware of mistakes you have been making, note them somewhere, and avoid them the next time.

I do, in fact, follow this practice. I have at least a dozen usage reference books in my library—some of them devoted exclusively to errors—and, once or twice a week, I open one of the the books anywhere and start to read. In this way, learning about usage is not nearly so difficult as you might expect. (Until last year, I might have written "not nearly *as* difficult" except that I happened to read that in this construction *as* is replaced with *so* after *not.*)

Before I go further, though, I must say a bit about *standards.* Just recently I told a group of computer people that the word *orientate* should not exist (even though it appears in the dictionary). I said that the correct word is *orient.* To which a very senior manager responded that I was a pedant, that everyone in his firm says *orientate* all the time, and that, therefore, it must be acceptable.

We each had a point. In a natural language, usage determines standards. If everyone believes that *presently* means "now" rather than "soon" (which is what I think it means), then eventually the dictionary will change the accepted definition. (Right now it is the fifth or sixth definition in some dictionaries; over the years it will rise on the charts like a popular song.)

Professional writers and editors like me tend to be elitists. We like to believe that "correct" standards are those accepted by the "right" people, by people with the right kind of education and sensitivities.

Whether you are a populist or an elitist, you should make it your policy to look up disputed usages and to be cautious before using a word in a way that is frowned on by the dictionary or reference book. Interestingly, there is only one dictionary I know of that reports the vote of its usage panel. On controversial items, *The American Heritage Dictionary of the English Language* tells you the vote of its expert panel. Thus, even though the opinion is expressed by an "elite," the dictionary abandons the notion that there is a single, authoritative ruling on these matters.

The *American Heritage* panel, for example, voted 47 percent in favor of *presently* in the sense of "now." But 72 percent voted against *anxious* as a synonym for *eager.* (Nearly all the panelists—95 percent—insist that *media* is plural, and 50 percent find *data* acceptable as a singular. Fully 84 percent of the panelists would not use *due to* at the beginning of a sentence!)

If you want to write well, you will take your advice from the experts, not from your colleagues. When there is an official style guide or style manual (sometimes cited in a contract), you will follow that source. The only times you may want to ignore the advice of the experts are when the experts are divided (as on *data*), or when doing it the proper way would sound absurd or odd to *your reader.* (I have never met an example of this last problem, but I suspect it happens occasionally.)

The most regrettable effect of this confusion about standards is that sometimes I am afraid that if I were to use a word correctly (like the *were* in this sentence), people might think I had made a mistake. Whenever a word or usage goes into this limbo of uncertainty, your best choice is to avoid it for a while.

With this background, then, here are some tips on usage.

Misused and Confused Most of us use only words whose meanings we know. In general, when we are not sure of the meaning of a term, we avoid it. There are, however, a few words that are chronically misused. For example:

WORD	WRONG MEANING	RIGHT MEANING
presently	now	soon
enormity	hugeness	evil
comprises	makes up	includes
embattled	harassed	ready to fight
infer	imply	deduce
hopefully	we hope	with hope
verbal	oral	in words
provision	providing	contract clause
anticipate	predict	act in advance
while	although	at the time

When in doubt, always consult a reliable dictionary; eventually these persistent mistakes of usage have a way of becoming acceptable—but not to demanding readers!

Most of our misused words are actually words that we have confused with some similar-sounding words. Be careful of these pairs of frequently confused terms:

CONTINUOUS means *all the time.*
CONTINUAL means *often and intermittent.*

AVERSE means *opposed to.*
ADVERSE means *unfriendly* or *dangerous.*

APPRISE means *inform.*
APPRAISE means *rate* or *evaluate.*

AFFECT (verb) means *influence.*
EFFECT (verb) means *cause* or *"effectuate."*

LIE (verb) means *rest* or *repose.* (lie, lay, lain)
LAY (verb) means *put* or *place.* (lay, laid, laid)

ALTERNATIVE means *one of several ways.* (originally one of two ways)
ALTERNATE means *substitute.*

MASTERFUL means *overpowering* or *dominant.*
MASTERLY means *skillful, artful.*

FORTUITOUS means *accidental, unplanned.*
FORTUNATE means *lucky, desirable.*

PERCENT means *a unit of measurement.*
PERCENTAGE means *proportion.*

Again, consult a dictionary if you are unsure.

Also refrain from inventing or coining new terms unless there is no English word that means what you are trying to say.

Several of these counterfeit coins are in wide circulation. Watch out for them:

COUNTERFEIT	REAL
remediate	remedy
administrate	administer
orientate	orient
irregardless	regardless
preventative (as an adjective)	preventive
firstly	first
attendee	participant
interviewee	respondent
strategical	strategic
deinstall	remove
deselect	reject

"Because" As I said under the discussion of complex sentences, there is a strange phobia in America about beginning a sentence with *because.*

(Many of my clients remember being told by teachers that *because* should not appear at the beginning of sentences.) Again, there simply is no such rule!

Instead of *because,* then, many people try unacceptable replacements: *since* (ambiguous), *as* (weak and too British), *due to* (not suitable for the beginning of a sentence), or *on account of* (correct but long-winded). We also have *due to the fact that, for the reason that, based on the conclusion that,* and even weirder possibilities.

NO:	*Due to* the strike our installation was delayed.
YES:	*Because of* the strike our installation was delayed.
NO:	*As* the update had not been completed, we delayed the quarterly report.
YES:	*Because* the update had not been completed . . .
NO:	The sponsoring agency was worried *since* we were over budget.
YES:	The sponsoring agency was worried *because* we were over budget.

Generally, the phrase *due to* should appear after the verb *to be.* (See *due to/prior to.*) If you want to use *since* in the sense of time, but fear it will be misread as *because,* use the phrase *ever since.*

"Due to/Prior to" The terms *due to* and *prior to* are controversial and troublesome. Do not use *due to* as a synonym for *because of,* and do not use *prior to* as a preposition meaning *before.*

Use *due to* only as a synonym for *attributable to* and only after the verb *to be (is, are, was, were, be)* or after some other "linking verb" (such as *seems, feels, appears*). Use *prior to* only as a synonym for *earlier than* and only after the verb *to be* or a "linking verb."

DUE TO

NO:	*Due to* the number of user mistakes, we are rewriting the manual.
YES:	*Because of* the number of user mistakes, we are . . .
	Or
YES:	The replacement of the manual is *due to* the number of user errors.

PRIOR TO

NO:	*Prior to* the demonstration for senior management, the middle managers attended a training program.
YES:	*Before* the demonstration for senior management . . .
	Or
YES:	The training program for middle managers was *prior to* the . . .

Many careful writers avoid *prior to* except in papers on logic; if you are unsure about *due to,* avoid it—and stop using it at the beginning of sentences and clauses.

"Not" *Not* is a troublesome word. Be especially careful of two problems:

- Confusing *not all* with *all not*
- Negating both parts of a compound (that is, negating two verbs when you intend to negate only the first)

NOT ALL

NO:	*All* of the old modules *do not fit* the new version.
YES:	*Not all* of the old modules *fit* the new version.

COMPOUND

NO:	New project managers *will not learn* from the mistakes of their predecessors and *make* all the same errors.
YES:	New project managers *will not learn* from the mistakes of their predecessors; instead, they *make* all the same errors.

Good writers avoid *not*—except when they are making a strong denial. Use *not* only when you believe it should be underscored for emphasis.

Troublesome Prepositions Many English idioms contain prepositions that are hard to keep straight—hard even for native speakers of the language.

Different *Different* sounds much better with FROM than with THAN.

Concur You concur WITH someone.
You concur IN an idea or assertion.

Disappoint You are disappointed IN someone.
You are disappointed WITH something.

Graduate *Graduate* takes the preposition FROM.

Adept *Adept* works better with IN than with AT.

Identical *Identical* works better with WITH than with TO.

Of course, when everyone stops using the "correct form" of these idioms, the incorrect form becomes acceptable. So many people say *different than* and *identical to* that these uses are nearly standard. A good writer, though, will want to use the safest conventions.

"Amount/Number" Certain words work with countable (digital) objects; others work with continuous (analog) volumes and quantities.

The most commonly misused pairs are *less/fewer, amount/number,* and *much/many.*

LESS/FEWER

NO:	We had *less* bugs in this program.
YES:	We had *fewer* bugs in this program.
	Or
YES:	We had *less* trouble with this program.

AMOUNT/NUMBER

NO:	The new process increased the *amount* of breakdowns.
YES:	The new process increased the *number* of breakdowns.
	Or
YES:	The new process increased the *amount* of downtime.

MUCH/MANY

NO:	*Much* of the changes were due to a new management.
YES:	*Many* of the changes were due to a new management.
	Or
YES:	*Much* of the improvement was due to a new management.

Modifiers: Misplaced and "Squinting" Certain modifiers are slippery; they slide into the wrong position in the sentence. The most dangerous are *only, nearly, almost, already, even,* and *just.*

ALMOST

NO:	They *almost* worked five years on that system.
YES:	They worked *almost* five years on that system.

ONLY

NO:	The data base *only* contains the year-end summaries for 1970–75.
YES:	The data base contains *only* the year-end summaries . . .
	Or
YES:	The data base contains the year-end summaries *only* for 1970–75.

NEARLY

NO:	The sponsor *nearly* wanted seventy-five changes in the report.
YES:	The sponsor wanted *nearly* seventy-five changes in the report.

JUST

NO:	The vital statistics system *just* reports the effects of poisoning incidents in terms of deaths.
YES:	The vital statistics system reports the effects of poisoning incidents *just* in terms of deaths.

In general, these slippery descriptors should appear just before the terms they modify.

Sometimes a modifier will be placed in such a way that we cannot tell which of two words it is modifying. We call such descriptors "squinting" because we do not know which way they are looking.

NO:	Senior analysts who talk about the past *constantly* confuse novice programmers.
YES:	Senior analysts who talk *constantly* about the past confuse novice programmers.
	Or
YES:	Senior analysts who talk about the past *are always confusing to* novice programmers.

Usually we are blind to our own squinting descriptors. They, among other problems, are almost always detected by someone else.

Optional Words When a word is optional, cut it—unless keeping it will make the sentence clearer or easier to read. The most common cases are (1) *that* at the beginning of a noun clause used as an object and (2) the first two words in a nonrestrictive clause.

"THAT" CLAUSES

The salesperson claimed (that) distributed processing would be more cost-effective than the current configuration.
Everyone in the industry believes (that) the RFP is wired.

NONRESTRICTIVE CLAUSES

Read the terms and conditions, (which are) described in Chapter 6.
The candidate, (who is) an MIT man, is overqualified for this position.

If there is any possibility that cutting the optional word will make the sentence harder to read, do not cut!

For example, in the sentence

John believed the doctor would arrive in time.

it is easy to misread the first four words ("John believed the doctor"). To avert that misreading, you would do best to include the optional *that,* as in

John believed *that* the doctor would arrive on time.

Grammar Bugs

Most adults know the grammar of their language. Actually, they do not "know" it so much as use it correctly by instinct.

Most of us can speak in sophisticated sentences that contain appositions and subordination long before we have any idea what "appositions" and "subordination" are. And many of us continue to speak and write in sophisticated sentences long after the grammar lessons of the seventh and eighth grades have faded from memory.

Grammar is mainly instinctive, learned at a time in our lives when what we learn is virtually wired into us permanently. That is why learning a second language is so difficult for adults, and also why incorrect grammar learned at one's mother's knee is so hard to shake. (A child who says "The reason is because . . ." when he or she is four years old is likely to keep on saying it forever.)

If English is your second language, you probably know more of its official rules and regulations than do the people for whom it is a first language. But you will inevitably be frustrated for years with hundreds of special rules, and exceptions to the special rules, that are not in your books and that even a native speaker of English could not explain to you. To an extent this is true of all languages: It takes many years of careful study and excellent coaching to become even competent in another language, and, even for the brightest students, sometimes the influences of one's first language are *never* gone.

English, though, is especially difficult in this regard, because it is the least systematic and consistent of the major languages. Its rules for word order (though easily mastered by little children) elude sophisticated computer programmers. Its variety and range and acquisitiveness—its tremendous vocabulary and boundless capacity for generating new terminology—distinguish it among languages. It is one of the most versatile and powerful languages—and one of the hardest to master!

Consequently, if English is your second language, you should ensure that there is always an editor (a friend or secretary or "real" editor) to check your idioms and syntax. (In some firms with a high proportion of people who speak English as their second or third language, the secretaries are hired for their editorial skills.)

If neither you, nor your closest colleague, nor the person you are married to, nor your supervisor speaks English as a first language, then please secure some editorial assistance right away!

But even if English is your first language, there is still a good chance that you are prone to certain errors (mainly those you acquired as a small child). Although it may prove embarrassing, you should seek out criticism of your grammar. Get someone to tell you your little mistakes (*try and* instead of *try to,* for example), and post them over your desk.

Each region of the country has its peculiar grammar, curious expressions that are acceptable in speech but barbaric in writing. Find someone who, as far as you're concerned, speaks a high-quality, "educated" brand of English and get him or her to point out your flaws. There may be a momentary embarrassment, but it will spare you greater embarrassment later on.

Most grammar bugs, then, are idiosyncratic: Each of us has specific ones. There are, however, a few grammar errors that I see often in my work. These errors, by the way, are not those made by ignorant or uneducated people; they are the ones made by the best writers, including the professionals.

Subject-Verb Agreement We all know that singular subjects take singular verbs and plural subjects take plural verbs. Often, though, we are misled by some noun between the subject and the verb.

NO:	An *estimate* of how long it will take to implement the full security requirements *are* beyond the scope of this report.
YES:	An *estimate* of how long it will take to implement the full security requirements *is* beyond the scope . . .
NO:	*Results* of studies on this system *has* had significant impact on the company's marketing strategy.
YES:	*Results* of studies on this system *have* had significant impact on the company's marketing strategy.
NO:	A more-detailed *discussion* of how to apply security and what security is available *are* contained later in this guide.
YES:	A more-detailed *discussion* of how to apply security and what security is available *is* contained . . .

Be especially careful of subjects that start with a singular noun but *seem* to become plural—*each of the engineers,* or *a group of physicians.* These subjects take *singular* verbs.

When the subject is *either* or *neither,* the verb is singular:

Either is acceptable.

Neither has worked.

Either of these approaches *is* acceptable.

Neither of their systems *has* worked.

When the subject of the sentence is an *either-or* phrase or a *neither-nor* phrase, the rule is a bit more complicated. If all the terms in the subject are singular, the verb is singular. If all the terms are plural, the verb is plural. If the terms are mixed, the number of the verb is decided by the term *closest to the verb* (the only case in English where the number of the verb is affected by the position of the subject).

YES:	Neither this *plan* nor your first *one is* adequate. (both singular)
YES:	Either your *calculations* or your *drawings are* incorrect. (both plural)
YES:	Neither your preliminary estimates nor your detailed *projection is* accurate. (singular closest)
YES:	Neither your detailed projection nor your preliminary *estimates are* accurate. (plural closest)

Pronoun Reference Every pronoun must point clearly to some noun or noun phrase. Pronouns without clear antecedents are treacherous for the reader.

HIS

NO: He told the controller *his* phone was not working.

WHOSE PHONE?

THEY

NO: Operator's manuals should not be given to new employees until *they* have been date-stamped.

WHO OR WHAT IS TO BE DATE-STAMPED?

IT

NO: The new maintenance policy offers no more service than the last, which cost 30 percent less. *It* provides liability insurance, though.

WHAT PROVIDES LIABILITY INSURANCE?

In the absence of a clear antecedent, a pronoun is presumed to refer to the closest eligible noun that precedes it. This "default" rule can produce some weird meanings.

Troublesome Plurals Take time to know when a word is singular, when plural. Do not use plural nouns with singular verbs; do not use stuffy Latin plurals when writing in English.

SINGULAR	PLURAL
medium	media
criterion	criteria
phenomenon	phenomena
formula	formulas
honorarium	honorariums
appendix	appendixes
agenda	agendas

Data is used about equally often as a singular or plural. No matter how you use it, use it consistently.

"Split Infinitives" Many people object to "splitting infinitives," that is, to placing a modifier between the *to* and the verb in the infinitive. They would

complain about such expressions as *to entirely separate* or *to be quickly lifted* (the passive form). Thus, split infinitives are dangerous for the amateur writer—even though there is really nothing wrong with them!

To correct a split infinitive, you may (1) remove the modifier, or (2) relocate the modifier, or (3) find a better verb that takes away the need for the modifier.

	REMOVE
NO:	I intend to *entirely* finish this report today.
YES:	I intend to finish this report today.
	RELOCATE
NO:	. . . to *boldly* go where no man has gone before . . .
YES:	. . . to go *boldly* where no man has gone before . . .
	REPLACE THE VERB
NO:	They have not had time to *totally learn* the language.
YES:	They have not had time to *master* the language.

Of course, if you really believe that the split form is the clearest and best form, you may decide to deliberately split it anyway. Remember, though, that there may be some fussy people among your readers who will regard your split infinitive with righteous contempt.

"Ending With A Preposition" Most people know that a preposition is a word you should not end a sentence with. Unfortunately, many English verbs have prepositions attached. And these attachments tend to appear at the ends of sentences and clauses.

If possible, choose some other verb when you discover that your first choice forces you to end with a preposition.

NO:	Here is a problem we should attend *to.*
AWKWARD:	Here is a problem to which we should attend.
FORMAL:	Let us attend to this problem.
BEST:	Here is a problem we should address.

Be careful of verbs like these:

- work on
- listen to
- open up
- put up with
- make a fool of
- grant the truth of

no one bothers to use *whom* in all the necessary places. At the very least, though, be sure to use *whom* when it comes right after a preposition: *for whom, to whom, by whom, with whom.* In these cases, *who* would be ghastly.

As much trouble as we have with case, we have even more with mood. Most of the people I meet have no idea of when to use *shall* and *will,* and some insist that *shall* need be used only in contracts and laws.

In fact, the rules on *shall* and *will* are simple. The term you use depends on two factors: first, the *person* of the subject, and, second, whether the statement is a plain remark about the future or a proclamation.

	STATEMENT	PROCLAMATION
I, We	shall	will
You	will	shall
He, She, It, They	will	shall

Most people mistakenly believe that *will* goes with all the statements and *shall* with all the proclamations, oaths, laws, commands, directives, and so forth. In fact, the reason that most laws and contracts say *shall* is that they are written in the second or third person (*you shall,* or *the contractor shall).* In the first person, a writer wanting to make a strong promise or proclamation would have to write *I will.* The general who said "I shall return" got it wrong.

Another small, but persistent, problem with verbs is the past subjunctive, the use of *were* in place of *was.* These three sentences are correct:

- I wish this contract *were* signed.
- Suppose this *were* the only technique appropriate.
- If he *were* project director, we would have finished under budget.

Again, these sentences are correct. If you were to replace the *were* with *was,* most editors would object. The rule is simple: When the statement is hypothetical or contrary to fact, use the subjunctive *were* instead of *was.*

Do not worry about the present subjunctive; it has nearly vanished from modern English. I do not know of anyone who would write

If this *be* acceptable . . .

Punctuation Bugs

There are three kinds of punctuation:

- Marks that you *must* use, to avoid making a mechanical error
- Marks that you *may* use, or leave out if you prefer
- Marks that are one of several alternatives you may *choose* from

There are relatively few "musts," and most writers have no trouble with them. You *must* put a period at the end of a simple declarative sentence, and you *must* put that period *inside* the quotation marks if that sentence ends with quotation marks.

In contrast, you may put a comma after short introductory words or phrases ("Later, we decided to try . . ."), even though many editors would remove it. (I tend to keep the optional commas in my work.)

And, in further contrast, you may have to choose among several ways of setting off an explanatory or appositive phrase. Here are some of the choices:

- The manager used no quantitative techniques, such as PERT.
- The manager used no quantitative techniques (such as PERT).
- The manager used no quantitative techniques—such as PERT.
- The manager used no quantitative techniques: such as PERT.

Writers have wide latitude in setting off "nonrestrictive" phrases and clauses, and the differences among these alternatives are usually insignificant.

The least-controversial part of any writer's handbook is the punctuation section. (There is much less dispute about semicolons than about split infinitives.) A good dictionary will include long entries for each of the punctuation marks and show you the DOs and DON'Ts. As you do your final editing, though, be alert to these few problems I have listed below. Again, I am pointing out the mistakes of people who are reasonably skillful as writers, not those of people who lack basic knowledge of English rules.

Commas Do not connect two sentences or independent clauses with a comma. (This error is called the "comma splice.")

NO:	No applicant is rejected because of age, this policy has been in effect for years.
YES:	No applicant is rejected because of age. This policy has been in effect for years.
	Or
YES:	No applicant is rejected because of age; this policy has been in effect for years.
	Or
YES:	No applicant is rejected because of age—this policy has been in effect for years. (informal, but acceptable)

Do not set off restrictive clauses with commas. That is, do not set off any clause that could not be removed from the sentence without distorting its meaning.

Hyphens Do not assume that when you add a prefix to a word you must put a hyphen in it. There are just a few cases where a hyphen is needed. And it is almost never needed when the prefix is *re-, pre-, sub-,* or *semi-*. Use a hyphen only when

1. The prefix ends in *a* or *i* and the base word starts with the same letter: anti-intellectual, ultra-adaptive.
2. The hyphen is needed to prevent mistaking the word you intend for some other word: reform vs. re-form, remark vs. re-mark, refuse vs. re-fuse.
3. The prefix is *self:* self-addressed, self-administered (but not selfsame or selfless).
4. The word is likely to be mispronounced without the hyphen: co-op, co-worker.
5. The base is capitalized: non-American, pro-Israel.
6. The base is a number: pre-1945; post-1978.

Note that there is no hyphen in *prerequisite, redesign, subset, infrastructure,* or *supercharged.*

Semicolons A semicolon is almost equivalent to a period. Grammatically, the semicolon is like an *,and.* Its beauty is that it allows nearly a full stop (like a period) while still managing to link two large sentences or clauses (like *,and*). Be sure, though, that the matter to the right of the semicolon is a *complete sentence.*

NO:	We wanted a senior project manager; that is, one who had installed projects on this scale.
YES:	We wanted a senior project manager, that is, one who had installed projects on this scale.
NO:	Our company has had thirty years' experience in software, we are the oldest in the business.
YES:	Our company has had thirty years' experience in software; we are the oldest in the business.
NO:	The plan lacked three ingredients; money, personnel, and willpower.
YES:	The plan lacked three ingredients: money, personnel, and willpower.
WEAK:	The specifications were odd, and the RFP looked wired.
STRONGER:	The specifications looked odd; the RFP looked wired.

Be careful not to overuse semicolons. Do not feel that just because two sentences are short, they must be joined with a semicolon. Join them only if they are closely related in meaning.

Colons Use colons to introduce a list or a long example. Use them to show that what follows the colon is an explanation or elaboration of what preceded the colon. Be careful, though. The material to the right of the colon may be either a complete sentence or a fragment. The material on the left, however, should be a self-contained statement or sentence.

NO:	We bought: two printers, four CRTs, and a reader.
YES:	We bought two printers, four CRTs, and a reader.
	Or
	We bought the following items: two printers, four CRTs . . .
NO:	Omega promised us a decision by Thursday: they later told us it would be Tuesday.
YES:	Omega promised us a decision by Thursday; they later told us it would be Tuesday.
	Or
YES:	Omega asked for another extension: to Tuesday.
WEAK:	He overlooked the system's most serious flaw, which was that the table headings were coded.
BETTER:	He overlooked the system's most serious flaw: The table headings were coded.

Lately, editors have become much more tolerant about using colons before an itemized list. In the past, if the material before the colon was only a fragment, most editors would prefer a dash to a colon.

	Traditional
NO:	The main advantages are:
YES:	The main advantages are—
YES:	The main advantages are
	Or
YES:	The main advantages are as follows:

Nowadays, though, lists are so popular (especially lists with "bullets") that colons are appearing in all sorts of places that would have been unacceptable only a few years ago.

Dashes Do not be afraid of the dash; it signals a pause slightly longer than a comma, or it interrupts the sentence for a purposeful digression.

1. Type dashes correctly. *Word--word.* Two hyphens--no spaces.
2. If you want to set off an apposition or an interjection within a sentence, be sure to use a dash at both ends. But no more than one pair of dashes to a sentence!
3. In less-formal messages (like memos), you may even use a dash to connect two closely related sentences. Use the dash instead of the semicolon—don't be afraid.

Dashes are becoming more acceptable all the time. Although they were once considered the favorite punctuation mark of the uneducated, they have become a useful way to provide breathing space in long or complicated sentences.

Checking for Conventions

If you are sure you have the right words (need I add, spelled correctly); if you are sure there are no grammar bugs; if you are sure that you have punctuated things correctly and intelligently . . . then about all that is left is to make sure that you have used your conventions consistently.

By *conventions* I mean all those choices of spelling, punctuation, and format for which there is no single right way. In which you choose to do it one way rather than another. (Is *data* singular or plural, for example? Is *DBMs* or *DBM's* the plural of *DBM?*)

As long as you have selected your conventions intelligently—consulting an appropriate style guide—your only concern should be to see that your conventions are *consistent.* Conventions that are not used the same way *every* time tell your reader that you are careless or lazy. They distract and irritate.

Even when a large team of authors works on one document, someone must be responsible for seeing that all conventions are used consistently.

The most efficient way to achieve this consistency, of course, is to have a style guide prepared for your company. If every typist and every document manager has a copy of the guide—and if they can be persuaded to use it—then the early drafts of the report or manual will already be consistent. (At the very least, every organization should have a correspondence manual so that the general style of every letter will be consistent.)

Whether or not you have such a guide, whether you have an editor or a skillful typist to help you, please pay especially close attention to these conventions.

1. Capitalization and Abbreviation: If a word is capitalized once, it should always be capitalized. Yet, I have seen short documents that contained *federal government, Federal government,* and *Federal Government.* I have also seen reports in which Health Maintenance Organizations were called, in the singular, HMO and H.M.O, in the plural, HMOs, HMO's, H.M.O.s, and H.M.O.'s.

There is no point in asking which of these is correct. The point is that they cannot all be correct in the same document.

2. Spelling: Certain words have two or more spellings. There are *enclose/inclose, employee/employe, judgment, judgement.* Use the *first* spelling in the dictionary, unless your company's style guide says otherwise.

3. Hyphens: Among the punctuation marks, hyphens are the most variable. As I said earlier in this chapter, many hyphenated words do not need them (*prerequisite* rather than *pre-requisite,* for example). But there are many cases, especially compound nouns and adjectives, where you have some choice. Which do you like: *decision-maker, decision maker,* or *decisionmaker?*

Often, your style guide will prefer one of these forms. But other times you are on your own. What about *data base* as a noun and *data-base* as an adjective? Once you choose, be consistent.

4. Embellishments: Be careful in your choice of emphases and highlights. To stress or set off a term, you may use *underscoring, italicizing* (by changing the typewriter element), *quotation marks, capitalizing the first letter, capitalizing all the letters.* Other times you can draw boxes or circles around key terms or print them in a bolder type.

These embellishments can be used with single words or short phrases. They can also be used with entire sentences and paragraphs. With longer passages you also have the option of *indenting, changing the type style,* or even *changing the color.*

Again, there is a tendency in early drafts to use every sort of embellishment. But in the first draft it is silly to worry about consistent embellishments. Before the document leaves your desk, someone should bring some order to make sure that consistent or parallel or comparable embellishments are used throughout.

(For book-length documents you will almost certainly need a professional editor to look after this complicated problem.)

5. Page Layout: Worry about the overall look of your pages. Insist on uniform margins, consistent indentations, a single type font (except when you change it deliberately), and uniform spacing between lines and paragraphs.

6. Lists and Numbers: When you put items into lists, use the same listing conventions. Do not switch from numbers to letters to bullets—especially in adjacent passages.

Be careful of numbers. Do not switch from Phase II to Phase 2. Do not call Figure IV-2 Figure 4-2. Whatever conventions you select for numbering items and exhibits, be careful and consistent.

In general, worry about any convention that you "made up." If you picked a peculiar or particular way to express or emphasize an idea, if you did it without referring to a standard guide or reference work, then make sure that your choice was legitimate, and that you followed it consistently throughout the document.

And if you find that you and your colleagues have recurring problems with the same conventions (how to number figures, for example), have someone in your organization prepare a guide and make sure that everyone uses it all the time.

THE FINAL CHECKLIST

Eventually, you have to let your message go. Of course, I want you to take as much time as possible to ensure that your report or letter or proposal is as nearly perfect as you can make it. But sooner or later you must say good-bye.

To satisfy yourself that it is time to release your work, go through this checklist.

Design and Organization

____ Did I know my exact objective for writing, and does it show?
____ Does the beginning capture the reader's attention?
____ Is my writing suitable for my audience? In tone, vocabulary, organization?
____ Is there a logic in my organization? Can the reader see it?
____ Have I told the truth?

Style and Language

____ Have I used words correctly and precisely?
____ Have I refrained from showing off?
____ Have I wasted words?
____ Have I built sentences that can be read easily?
____ Have I varied the length and form of my sentences?
____ Are my paragraphs coherent?

Refinements

____ Is the document neat and attractive?
____ Is the document free from all typing and transcription errors?
____ Is everything spelled correctly?
____ Are there errors of grammar? Punctuation?
____ Are all the conventions consistent?

(And, finally, the single most important question of all:)

____ Am I willing to bet my career and future income on this piece of work, exactly as it looks right now?

Unless you can give a resounding yes to this last question, then stop, stay a while.

In most scientific and technical enterprises, the quality of your writing will count more than its timeliness. There are exceptions, of course: proposals with fixed deadlines, final reports with harsh penalties for delays. Usually, though, the extra hour's or day's delay will hurt you far less than the careless or clumsy work.

Think of your written products as extensions of yourself. Take pride in your work.

Appendix: References and Resources

BOOKS AND JOURNALS ABOUT EFFECTIVE WRITING

There are scores of books that can teach you to write more effectively. Here are a few of the best:

Barzun, Jacques, *Simple & Direct.* New York: Harper & Row, Pub., 1976.

BROGAN, JOHN, *Clear Technical Writing.* New York: McGraw-Hill, 1973. (Note: Despite its title, Brogan's book is for *anyone* who wants to write lighter, more-readable sentences. It is the best medicine I know for wordiness.)

EWING, DAVID, *Writing for Results in Business, Government, the Sciences and the Professions* (2nd ed.). New York: John Wiley, 1979.

STRUNK, WILLIAM, and E. B. WHITE, *The Elements of Style* (2nd ed.). New York: Macmillan, 1972. (Note: A masterpiece. No home should be without one.)

TICHY, HENRIETTA, *Effective Writing for Engineers, Managers, and Scientists.* New York: John Wiley, 1966.

WADDELL, MARIE, ROBERT ESCH, and ROBERTA WALKER, *The Art of Styling Sentences.* Woodbury, N.Y.: Barron's Educational Series, 1972.

For those who take a certain pleasure in reading about writing, especially if they like to see pretentious writing ridiculed, try these:

MITCHELL, RICHARD, *Less Than Words Can Say.* Boston: Little, Brown, 1979.

NEWMAN, EDWIN, *On Language.* New York: Warner Books, 1980.

For those who need something simpler, try an excellent book for young readers:

RUBINS, DIANE T., *The A+ Guide to Good Writing.* Englewood Cliffs, N.J.: Scholastic Book Services, 1980.

Among the many journals devoted to better writing in general, try these:

College English

IEEE Transactions on Professional Communication

HANDBOOKS AND RULEBOOKS

The best little handbook of grammar and mechanics I have seen is, indeed, little:

CORBETT, EDWARD P.J., *The Little English Handbook: Choices and Conventions.* New York: John Wiley, 1973.

There is also a host of handbooks meant for college freshmen. Typical among them is

IRMSCHER, WILLIAM, *The Holt Guide to English.* New York: Holt, Rinehart & Winston, 1972.

Many of the newer handbooks are quite tolerant of current usage. They usually say "most people prefer" instead of "always be sure to." Among the best of these more-liberal guides is

SABIN, WILLIAM A., *The Gregg Reference Manual* (5th ed.). New York: McGraw-Hill, 1977.

Most of the major scientific organizations have their own "style guide" or "style sheet"; be sure you get the one used by members of your profession. And if you work for the federal government (directly or under contract), get

The U.S. Government Printing Office Style Manual (rev. ed.), 1973.

For the mechanics and etiquette of business letters, consult any of the accessible secretarial guides. Among the easiest to find is

Webster's Secretarial Handbook. New York: G & C Merriam Company.

And for those of you who have simply forgotten your basic grammar, and who cannot pick it up from the handbook, I suggest

Learning Technology Inc., *Writing Skills 1 and 2.* New York: McGraw-Hill, 1970. (Note: These are programmed texts that lead you through the subject one small step at a time.)

USAGE

The best reference book on English usage is a good, recent dictionary. Because of the quality of its usage notes, I recommend

The American Heritage Dictionary of the English Language, New College Edition. New York: Houghton Mifflin, 1978.

To supplement your dictionary, acquire one or more of the following:

BERNSTEIN, THEODORE, *The Careful Writer.* New York: Atheneum, 1965.

______, *Reverse Dictionary.* New York: Quadrangle, 1977.

BERRY, THOMAS E., *The Most Common Mistakes in English Usage.* New York: McGraw-Hill, 1971.

FOLLET, WILSON, *Modern American Usage,* ed. Jacques Barzun. New York: Hill & Wang, 1966.

SHAW, HARRY, *Errors in English and Ways to Correct Them* (2nd ed.). New York: Barnes & Noble, 1970.

TECHNICAL WRITING AND REPORTING

There are too many books about technical writing to list them all. Typical among them are

JORDAN, S., and others, *Handbook of Technical Writing Practices.* New York: Wiley-Interscience, 1971.

MILLS, GORDON, and JOHN WALTER, *Technical Writing* (3rd ed.). New York: Holt, Rinehart & Winston, 1970.

NORGARD, MARGARET, *A Technical Writer's Handbook.* New York: Harper & Row, Pub., 1959.

RATHBONE, R. R., *Communicating Technical Information.* New York: Addison-Wesley, 1966.

SHERMAN, THEODORE, and SIMON JOHNSON, *Modern Technical Writing* (3rd ed.). Englewood Cliffs, N.J.: Prentice-Hall, 1975.

SMITH, RICHARD W., *Technical Writing.* New York: Barnes & Noble, 1963.

ULMAN, J. N., and J. R. GOULD, *Technical Reporting.* New York: Holt, Rinehart, & Winston, 1972.

WEISMAN, HERMAN, *Technical Report Writing.* Columbus, Ohio: Merrill, 1966.

The best way to get ideas that will improve your technical communication is to read some of the many journals on the subject. Among the best are

Journal of Technical Writing and Communication

Medical Communications

Technical Communication

The Technical Writing Teacher

PROPOSALS

Many of the books and journals listed above contain sections about proposals. And almost everything that applies to good report writing applies to proposals as well. Still, there is too little in print about the craft of proposals—and much of it is in hard-to-get, expensive publications.

Here are a few titles to get you started:

AMMON-WEXLER, JILL, and CATHERINE CARMEL, *How to Create a Winning Proposal.* Mercury Communications Corporation, 1976.

LEFFERTS, ROBERT, *Getting a Grant.* Englewood Cliffs, N.J.: Prentice-Hall, 1978.

Society for Technical Communication, *Proposals . . . And Their Preparation.* The Society, 1973.

TRACEY, J. R., D. E. RUGH, and W. S. STARKEY, *STOP: Sequential Thematic Organization of Publications.* Hughes Aircraft Corporation, 1965.

PAPERS AND ARTICLES

If you want to write better, more-publishable papers, you can learn nearly everything you need to know from

DAY, ROBERT A., *How to Write and Publish a Scientific Paper.* Institute for Scientific Information, 1979.

If you want to write interesting, readable articles for magazines, try

ZINSSER, WILLIAM, *On Writing Well* (2nd ed.). New York: Harper & Row, Pub., 1980.

If you want to pursue magazine writing more earnestly, try these monthlies:

The Writer

Writer's Digest

BUSINESS COMMUNICATIONS

There are also countless books on business writing, with the emphasis on letters. Here is a representative sample:

BLUMENTHAL, LASSOR, *Successful Business Writing.* New York: Grosset & Dunlap, 1976.

ELLENBOGEN, ABRAHAM, *Letter Perfect.* New York: Macmillan, 1978.

GOELLER, CARL, *Writing to Communicate.* New York: New American Library, 1974.

JANIS, HAROLD, and HOWARD DRESSNER, *Business Writing.* New York: Barnes & Noble, 1972.

VARDAMAN, GEORGE, and PATRICIA BLACK VARDAMAN, *Communication in Modern Organizations.* New York: John Wiley, 1973.

And an especially interesting periodical is

The Journal of Business Communication

Index

A

B

F

G

H

I

J

L

M

N

O

P

Q

R

S

T